Advances in Intelligent Systems and Computing

Volume 1157

The series "Advances in Intelligent Systems and Computing" contains publications on theory, applications, and design methods of Intelligent Systems and Intelligent Computing. Virtually all disciplines such as engineering, natural sciences, computer and information science, ICT, economics, business, e-commerce, environment, healthcare, life science are covered. The list of topics spans all the areas of modern intelligent systems and computing such as: computational intelligence, soft computing including neural networks, fuzzy systems, evolutionary computing and the fusion of these paradigms, social intelligence, ambient intelligence, computational neuroscience, artificial life, virtual worlds and society, cognitive science and systems, Perception and Vision, DNA and immune based systems, self-organizing and adaptive systems, e-Learning and teaching, human-centered and human-centric computing, recommender systems, intelligent control, robotics and mechatronics including human-machine teaming, knowledge-based paradigms, learning paradigms, machine ethics, intelligent data analysis, knowledge management, intelligent agents, intelligent decision making and support, intelligent network security, trust management, interactive entertainment, Web intelligence and multimedia.

The publications within "Advances in Intelligent Systems and Computing" are primarily proceedings of important conferences, symposia and congresses. They cover significant recent developments in the field, both of a foundational and applicable character. An important characteristic feature of the series is the short publication time and world-wide distribution. This permits a rapid and broad dissemination of research results.

** Indexing: The books of this series are submitted to ISI Proceedings, EI-Compendex, DBLP, SCOPUS, Google Scholar and Springerlink **

More information about this series at http://www.springer.com/series/11156

J. K. Mandal · Soumen Banerjee
Editors

Intelligent Computing: Image Processing Based Applications

 Springer

Editors
J. K. Mandal
Department of Computer Science
and Engineering
University of Kalyani
Kalyani, West Bengal, India

Soumen Banerjee
Department of Electronics
and Communication Engineering
University of Engineering and Management
Kolkata, West Bengal, India

ISSN 2194-5357   ISSN 2194-5365   (electronic)
Advances in Intelligent Systems and Computing
ISBN 978-981-15-4287-9   ISBN 978-981-15-4288-6   (eBook)
https://doi.org/10.1007/978-981-15-4288-6

This Springer imprint is published by the registered company Springer Nature Singapore Pte Ltd.
The registered company address is: 152 Beach Road, #21-01/04 Gateway East, Singapore 189721, Singapore

Preface

This volume entitled *Intelligent Computing: Image Processing Based Applications* contains ten research papers out of which a few papers are an extended versions of the 5th International Conference on Opto-Electronics and Applied Optics (OPTRONIX 2019) held from March 18 to 20, 2019, at the University of Engineering and Management, Kolkata, India, and the rest are taken from open circulations containing recent advancements in intelligent computing and development of machine learning and deep learning facilitated image processing applications in a variety of real-life applications including biomedical image analysis, intelligent security systems, real-time object detection, face detection, edge detection and segmentation, image captioning, content-based image retrieval. These are the development of new intelligent computing-based image processing applications, both in research problems and in real-world problems.

This book consists of contributions from various aspects of analysis, design and implementation of intelligent computing-supported image processing applications which are the fundamentals of associated fields/concepts, applications, algorithms and case studies.

The book contains 10 chapters. Chapter A Review of Object Detection Models Based on Convolutional Neural Network reviews some popular state-of-the-art object detection models based on CNN. In Chap. Cancerous Cell Detection Using Affine Transformation with Higher Accuracy and Sensitivity, cancerous cells are detected by affine transformation by working on multi-resolution, multi-lateral, multi-cluster image patterns for prediction analysis. In Chap. A Comparative Study of Different Feature Descriptors for Video-Based Human Action Recognition, a comparative study of different feature descriptors applied for HAR on video datasets is presented. Chapter Pose Registration of 3D Face Images deals with 3D face images, and Chap. Image Denoising Using Generative Adversarial Network discusses the way GANs are applied in the area of image de-noising. Chapter Deep Learning-Based Lossless Audio Encoder (DLLAE) addresses the deep learning-based audio encoding. PVD-based steganography, low back pain management, dermatologic classifications and retinal disease detection are the contents of the rest of the chapters.

Each paper was reviewed by at least two domain experts, and the accepted papers have been revised as per the norms of Springer AISC. Based on the comments of the reviewers, the papers were modified by the authors and the modified papers were examined for the incorporation of the same. Finally, ten papers are selected for publication in this special issue intelligent computing: image processing-based applications.

On behalf of the editors, we would like to express our sincere gratitude to the reviewers for reviewing the papers. Our sincere gratitude to the authority of Springer for giving us the opportunity to edit this volume.

Hope this special issue will be an effective material on the state-of-the-art information on the recent trends in image processing for the budding engineers and researchers.

Kalyani, India   Prof. J. K. Mandal
Kolkata, India   Soumen Banerjee

Contents

About the Editors

J. K. Mandal M. Tech. in Computer Science from the University of Calcutta in 1987, was awarded Ph.D. (Engineering) in Computer Science and Engineering by Jadavpur University in 2000. Presently, he is working as a Professor of Computer Science & Engineering and was former Dean, Faculty of Engineering, Technology & Management, KU, for two consecutive terms during 2008-2012. He was Director, IQAC, Kalyani University, and Chairman, CIRM and Placement Cell. He served as a Professor of Computer Applications, Kalyani Government Engineering College, for two years, as an Associate Professor of Computer Science for eight years at North Bengal University, as an Assistant Professor of Computer Science, North Bengal University, for seven years, and as a Lecturer at NERIST, Itanagar, for one year. He has 32 years of teaching and research experience in coding theory, data and network security and authentication, remote sensing & GIS-based applications, data compression, error correction, visual cryptography and steganography. He awarded 23 Ph.D. degrees, one submitted and 8 are pursuing. He has supervised 03 M.Phil., more than 70 M.Tech. and more than 125 M.C.A. Dissertations. He was Guest Editor of MST Journal (SCI indexed) of Springer. He has published more than 400 research articles out of which 167 articles in international journals. He has published 7 books from LAP Germany, IGI Global, etc. He has organized 31 international conferences and was one of the corresponding editors of edited volumes and conference publications of Springer, IEEE, Elsevier, etc., and he has edited 32 volumes as volume Editor.

Prof. Soumen Banerjee is Head of the Department of Electronics & Communication Engineering at the University of Engineering & Management, Kolkata, West Bengal, India. He was the visiting Faculty at the Department of Applied Physics, University of Calcutta, and an Ex-Faculty of IEM, Salt Lake, Kolkata. He is into the teaching profession for more than 18 years and received Best Teacher Award from UEM-Kolkata in 2018. He obtained his B.Sc. (Honors) degree in Physics, B.Tech. and M.Tech. degrees in Radio Physics and Electronics from the University of Calcutta. He is presently associated with research activities at the Indian Institute of Engineering Science and Technology (IIEST), Shibpur, India. His research interests include design, fabrication and characterization of wide band gap semiconductor-based Impatt diodes at D-band, W-band and THz frequencies, Substrate Integrated Waveguide (SIW) technology-based antennas, printed antennas and arrays. He has published more than 80 contributory papers in journals and international conferences and received Best Paper Award in IEEE OPTRONIX-2016 conference. He is the recipient of the National Scholarship, 2001 from Government of India, Ministry of HRD. Prof. Banerjee is the author of the books titled "Principles of Communication Engineering," "Electromagnetic Field Theory," "Communication Engineering-II," "Electromagnetic Theory & Transmission Lines," "Antenna Theory" and some others. He is a Corporate member of IEEE & IEEE AP Society and member of other professional societies like IETE (New Delhi) and IE (India). Prof. Banerjee is the reviewer of several Books Publishing Houses and also reviewer of several international journals like IEEE Sensors Letter, Microwave and Optical Technology Letters (MOTL-Wiley), Journal of Electromagnetic Waves and Applications (Taylor & Francis), Radioengineering Journal (Czech and Slovak Technical Univ.), Journal of Computational Electronics (Springer Nature), Journal of Renewable and Sustainable Energy (American Institute of Physics-AIP), and Journal of Infrared, Millimeter, and Terahertz Waves (Springer Nature).

A Review of Object Detection Models Based on Convolutional Neural Network

F. Sultana, A. Sufian, and P. Dutta

Abstract Convolutional neural network (CNN) has turned to be the state of the art for object detection task of computer vision. In this chapter, we have reviewed some popular state-of-the-art object detection models based on CNN. We have made a categorization of those detection models according to two different approaches: two-stage approach and one-stage approach. Herein, we have explored different gradual developments in two-stage object detection models from R-CNN to latest mask R-CNN as well as in one-stage detectors from YOLO to RefineDet. Along with the architectural description of different models, we have focused on training details of each model. We have also drawn a comparison among those models.

Keywords Convolutional neural network · Deep learning · Mask R-CNN · Object detection · RetinaNet · Review · YOLO

1 Introduction

In the field of computer vision, the task of predicting the class and the location of different objects contained within an image is known as object detection task. Unlike classification, every instance of objects is detected in the task of object detection. So object detection is basically instance-wise vision task. Before the popularity of deep learning in computer vision, object detection was done by using hand-crafted

F. Sultana · A. Sufian (✉)
Department of Computer Science, University of Gour Banga,
Mokdumpur, Malda, WB, India
e-mail: sufian.csa@gmail.com

F. Sultana
e-mail: sfarhana@ieee.org

P. Dutta
Department of Computer and System Sciences,
Visva-Bharati University,Santiniketan, WB, India
e-mail: paramartha.dutta@gmail.com

© Springer Nature Singapore Pte Ltd. 2020
J. K. Mandal and S. Banerjee (eds.), *Intelligent Computing: Image Processing Based Applications*, Advances in Intelligent Systems and Computing 1157,
https://doi.org/10.1007/978-981-15-4288-6_1

1

machine learning features such as scale invariant feature transform (SHIFT) [1], histogram of oriented gradients (HOG) [2]. At that era, the best object detection models gain around 35% mean average precision (mAP) on PASCAL VOC Challenge dataset [3]. But during 2010–2012, the progress in object detection has been quite motionless. The best-performing models gained their result building complex ensemble systems and applying few modification in successful models. Also, the representation of SHIFT and HOG could only associate with the complex cells in V1 of primary visual cortex. But, both the structure of visual cortex of animals and the object recognition process in computer vision found the fact that visual recognition might be a hierarchical, multilevel process of computing informative visual features. Inspired by this idea, Kunihiko Fukushima built "neocognitron" [4] which is a hierarchical, shift-invariant multi-layered neural network for visual pattern recognition. In 1990, LeCun et al. extended the theory of "neocognitron" and introduced the first practical image classification model based on convolutional neural network called "LeNet-5" [5, 6]. This model was trained using supervised learning through stochastic gradient descent (SGD) [7] via backpropagation (BP) algorithm [8]. After that, the progress of CNN was stagnant for several years [9]. In 2012, the resurgence of "AlexNet" [10] inspired researchers of different fields of computer vision such as image classification, localization, object detection, segmentation, and we have got many state-of-the-art classification models explored in [11] and object detection models like R-CNN [12], SPP-net [13], Fast R-CNN [14], YOLO [15], RetinaNet [16], SqueezeDet [17], StairNet [18], FRD-CNN [19] in successive years.

In this chapter, we have tried to give a review of different state-of-the-art object detection models based on convolutional neural network (CNN). We have described network architecture along with training details of various models in Sect. 2 with some contextual subsections. In Sect. 3, we have compared the performance of those object detection models on different datasets. Finally, we have concluded our chapter in Sect. 4.

2 Different Object Detection Models

After 2012, different researchers applied various strategies such as object proposal algorithm, pooling methods, novel loss functions, bounding box estimation methods along with convolutional neural network for betterment of the task of object detection. State-of-the-art object detection models based on CNN can be categorized into two different categories: (a) two-stage approach and (b) one-stage approach.

2.1 Two-Stage Approach

In the two-stage object detection, the first stage generates region or object proposals, and in the second stage, those proposals are classified and detected using bounding boxes (Fig. 1).

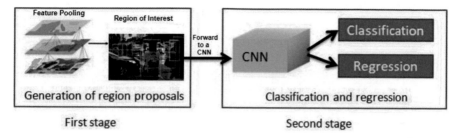

Fig. 1 Two-stage approach. Source: Authors

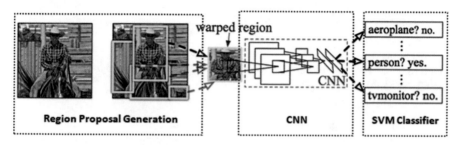

Fig. 2 Architecture of R-CNN [12]

R-CNN: Region proposal with convolutional neural network (R-CNN) [12] is the first CNN-based two-stage object detection model. The architecture of R-CNN contains three different blocks as shown in Fig. 2. First stage includes the first block where the authors have used selective search [20] algorithm to generate around 2000 class-independent region proposals from each input image. In the second block, following the architecture of AlexNet [10], they have used a CNN of five convolutional (conv) layers and two fully connected (FC) layers to pull out fixed length feature vector from each region proposal. As CNN requires fixed-sized image as input, the authors have used affine image warping [21] to get fixed-sized input image from each region proposals regardless of their size or aspect ratio. Then, those warped images are fed into individual CNN to extract fixed length feature vectors from each region proposal. The third block classifies each region proposal with category-specific linear support vector machine (SVM) [22]. Second and third blocks together constitute the second stage.

Training details: The authors have pre-trained the CNN of their model using ILSVRC-2012 [23, 24] dataset. Then, they have changed the ImageNet specific 1000-way softmax classifier with a 21-way classifier (for the 20 PASCAL VOC classes and one background class) and trained the CNN parameters using SGD with warped region proposals taken from PASCAL VOC [3] images. If the IoU overlap of region proposals with corresponding ground-truth box is ≥ 0.5, they treat the proposal as positive for that box class and rest as negative proposal. The learning rate of SGD was 0.001. The mini-batch size per SGD iteration was 128 (32 positive foreground

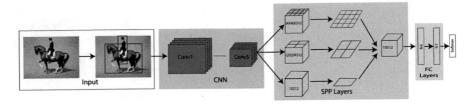

Fig. 3 Architecture of SPP-net

and 96 negative background window). They have optimized per class linear SVM using standard hard negative mining [25] to reduce memory consumption.

The use of CNN for object detection helped R-CNN to gain higher accuracy in detection problem, but there is a major downside of this model. CNNs need fixed-sized input image, but region proposals generated by R-CNN are arbitrary. To fulfill the CNNs requirement, the scale or aspect ratio or the originality of an image are got compromised due to cropping, warping, predefined scales, etc.

SPP-net: In a CNN, convolution (conv) layers actually do not need fixed size image as input. But fully connected (FC) layers require fixed length feature vectors as input. On the other hand, spatial pyramid pooling (SPP) [26, 27] can generate fixed-sized output regardless of input size/scale/aspect ratio. So, He et al. included SPP layer in between conv layers and FC layers of R-CNN and introduced SPP-net [13] as shown in Fig. 3. SPP layers pool the features and generate fixed length output from variable length region proposals, which are fed into FC layers for further processing. In this way, SPP-net made it possible to train and test the model with image of varying sizes and scales. Thus, it increases scale invariance as well as reduces overfitting.

Training details: The authors have followed [12, 28] to train their SVM classifier. They have also used ground-truth box to generate positive proposals. If the IoU of ground-truth box and positive window is ≤ 0.3, then they considered that window as negative sample. If negative sample—negative sample overlap—is ≥ 0.7, then that sample is removed. They have also applied standard hard negative mining [29] to train SVM. At test time, the classifier associates scores to the candidate windows. After that, the authors used non-maximum suppression [29] with threshold of 0.3 on the scored candidate windows. The authors have also followed [12] to fine-tune only the FC layers of their model. The model was trained using SGD. In each mini-batch, the ratio of positive and negative samples was 1:3. They started training 250k mini-batches with the learning rate 0.0001, and then 50k mini-batches with 0.00001. Also following [12], they have used bounding box regression on the pooled features from fifth convolutional layer. If IoU of a window and a ground-truth box is >0.5, then that is used in bounding box regression training.

As SPP-net uses CNN once on the entire image and region proposals are extracted from last conv feature map, the network is much faster than R-CNN.

Fast R-CNN: R-CNN and SPP-net had some common problems: (i) multi-stage pipeline training (feature extraction, network fine-tuning, SVM training, bounding box regression), (ii) training complexity in terms of space and time is high and also (iii) low object detection speed. Fast R-CNN [14] tries to overcome all the above limitations. Ross Girshick modified R-CNN and proposed Fast R-CNN which is a single-stage training algorithm that classifies region proposals and rectifies their spatial locations simultaneously. Fast R-CNN can train very deep detection network like VGG-16 [30] 9× faster than R-CNN and 3× faster than SPP-net.

R-CNN used CNN for each generated region proposals for further detection process. But, Fast R-CNN takes a whole image along with a set of object proposals together as input. From the extracted CNN feature maps, region of interests (RoI) are identified using selective search method. Then, the authors have used a RoI pooling layer to reshape the RoIs into a fixed length feature vector. After that, FC layers take those feature vectors as input and passes the output to two sibling output branch. One branch is for classification and another for bounding box regression. Figure 4 demonstrates the architecture of Fast R-CNN.

Training details: The authors have experimented their model using three types of network which are pre-trained on ImageNet [23] dataset. Three networks are CaffeNet AlexNet [10], VGG_CNN_M_1024 [31] and VGG16 [30] models. The networks have five max-pooling layers in total, and the number of conv layers in those networks was in between five to thirteen. During initialization of those pre-trained networks, the authors have first replaced the last max-pooling layer with a RoI pooling layer. Then, last fully connected layer and softmax layer of the networks were replaced with two sibling output layers for classification and bounding box regression. The authors have made minor alteration in the networks to take two inputs: a whole input image and a collection of RoIs present in those images. The authors have trained Fast R-CNN using stochastic gradient descent (SGD) with hierarchically sampled mini-batches. First, the authors have chosen N number of image samples randomly. Then, from each image, R/N number of RoIs are sampled for each mini-batch. During fine-tuning, they have chosen $N = 2$ and $R = 128$ to construct mini-batch. Initial learning rate for 30k mini-batch iterations was 0.001, and for next 10k mini-batch iterations, it was 0.0001. They have also used a momentum of 0.9 and parameter decay of 0.0005. They have used multi-task loss to jointly train their network for

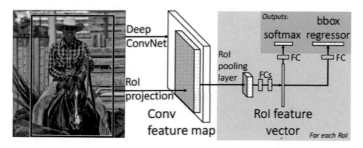

Fig. 4 Architecture of Fast R-CNN [14]

classification and bounding box regression in a single stage. They have followed [13, 14] to choose the ratio of positive and negative RoIs for training their model.

Using CNN once for an entire image and single-stage training procedure reduces the time complexity of Fast R-CNN in a large scale than R-CNN and SPP-net. Also, use of RoI pooling layer and some tricks in training helped their model to achieve higher accuracy.

Faster R-CNN: R-CNN, SPP-net and Fast R-CNN depend on the region proposal algorithm for object detection. All three models experienced that the computation of region proposals is time-consuming which affects the overall performance of the network. Ren et al. proposed Faster R-CNN [32], where they have replaced previously mentioned region proposal method with region proposal network (RPN). An RPN is a fully convolutional network (FCN) [33] which takes an image of arbitrary size as input and produces a set of rectangular candidate object proposals. Each object proposal is associated with an objectness score to detect whether the proposal contains an object or not (Fig. 5).

Like Fast R-CNN, the entire image is provided to the conv layers as an input of Faster R-CNN to produce feature map. Then, instead of using selective search algorithm, a RPN is used to identify the region proposals from the feature map. In RPN, the region proposals detected per sliding window locations are called anchors. An relevant anchor box is selected by applying a threshold value over the "objectness" score. Selected anchor boxes and the feature maps computed by the initial CNN model together are fed to RoI pooling layer for reshaping and the output of RoI pooling layer fed into FC layers for final classification and bounding box regression.

Training details: The RPN is basically a fully convolutional neural network [34] pre-trained on the ImageNet dataset, and it is fine-tuned using PASCAL VOC dataset. Region proposals, generated from RPN, along with anchor boxes are used to train the later part of Faster R-CNN after RPN. To train RPN, each anchor is assigned to a binary class label. Positive label is assigned to two types of anchors: "(i) the anchor/anchors with the highest Intersection-over-Union (IoU) overlap with a ground-truth box, or (ii) an anchor that has an IoU overlap >0.7 with any ground-

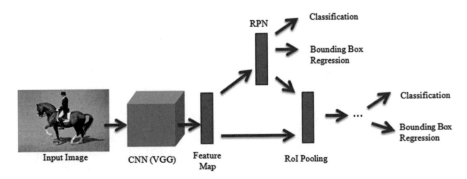

Fig. 5 Architecture of Faster R-CNN. Source: Authors

truth box." A negative label is assigned to an anchor if its IoU value is <0.3 for all ground-truth boxes. The authors have trained both RPN and Fast R-CNN independently. They simply followed the multi-task loss of Fast R-CNN to train their network. RPN is end-to-end trainable with backpropagation and SGD. The authors have also followed the 'image centric' sampling strategy of Fast R-CNN. A mini-batch of SGD is constructed with a number of positive and negative anchor boxes predicted from an input image. In each mini-batch, the ratio of positive and negative anchor could have a ratio up to 1:1. They have used a initial learning rate of 0.001 for first 60k mini-batches and 0.0001 for the rest 20k mini-batches. The authors have used 0.9 as momentum and 0.0005 as weight decay.

Previously mentioned object detector models considered single-scale feature map for object detection. In this way, those models lack exact location of object instances as the last layer feature map of CNN is scale invariant.

FPN: Lin et al. used the multilevel feature map of CNN in a different way to construct feature pyramid network (FPN) [35] for better object detection than state-of-the-art models. In Fig. 6, the bottom-up and top-down architecture of FPN is shown. The bottom-up pathway is the backbone CNN as used in Fast R-CNN. The top-down pathway consists a multilevel feature pyramid. Each level of top-down feature pyramid takes the upper layer's feature maps and the feature maps from corresponding stage of bottom-up pathway as input. Then, concatenate those feature maps to produce the output. FPN also used RPN to generate region proposals. On the region proposal, FPN is used for feature extraction. RoIs are predicted from different levels to fuse the property of multi-scale, and then, those RoIs are reshaped using RoI pooling layer as Fast R-CNN. The output of RoI pooling layer is fed into fully connected layer for classification and bounding box regression.

Training details: The authors used ResNet-based Fast R-CNN as the backbone network of FPN. Their network was end-to-end trainable using backpropagation. They have used synchronized SGD with mini-batch size of 2 images and 256 anchors per

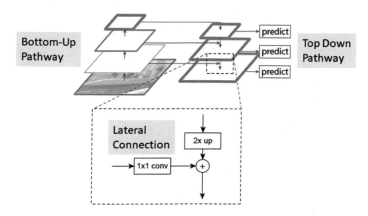

Fig. 6 Bottom-up and top-down architecture of FPN with a building block [35]

image. For first 30k mini-batches, the learning rate was 0.02 and 0.002 for next 10k mini-batches. The values of momentum and weight decay are 0.9 and 0.0001. The authors have trained their model using MS COCO dataset [36] with 80 categories.

Mask R-CNN: He et al. extended previous R-CNN object detection techniques to go one step further and locate exact pixels of each object instance (instance segmentation [37, 38]) instead of just bounding boxes. As this model masks each instances of an object independently, they named it Mask R-CNN [39]. Mask R-CNN has the exact region proposal network (RPN) from Faster R-CNN to produce region proposals. The authors applied *RoI Align* layer on the region proposals instead of RoI pooling layer to align the extracted features with the input location of an object. The aligned RoIs are then fed into the last section of the Mask R-CNN to generate three output: a class label, a bounding box offset and a binary object mask. For masking, a small fully convolutional neural network is used on each RoI. The architecture of Mask R-CNN is shown in Fig. 7.

Training details: The authors have used ResNet [40] and ResNeXt [41] networks as the base CNN of Faster R-CNN. Then, feature pyramid network(FPN) [35] is used with Faster R-CNN as backbone. Region of interests is extracted from different levels of FPN with varying scales. Use of ResNet and FPN helped Mask R-CNN to gain better accuracy and speed. The authors set all hyperparameters of their model following [14, 32]. If IoU of an RoI with a ground-truth box is ≥ 0.5, then that RoI is considered as positive, otherwise negative. Mask R-CNN also followed image-centric training. Mask R-CNN used multi-task loss function which combines the loss function of classification, localization and segmentation mask. The mask loss is defined only on positive RoIs. The ResNet_FPN network is trained using SGD with a mini-batch size of two images. Each image is sampled into N RoIs. The ratio of positive and negative RoIs for each mini-batch was 1:3. The learning rate for 160k iteration was 0.02. It was decreased by a factor of 10 for next 120k iteration. Also, the momentum and weight decay were 0.9 and 0.0001, respectively. Training of ResNeXt_FPN includes mini-batch size of 1 image and a initial learning rate of 0.01.

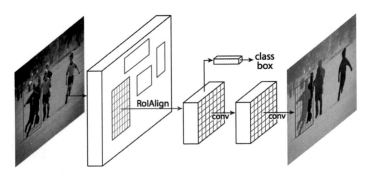

Fig. 7 Mask R-CNN [39]

2.2 One-Stage Approach

In the one-stage approach, classification and regression is done in a single shot using regular and dense sampling with respect to locations, scales and aspect ratio.

YOLO: The You Only Look Once (YOLO) [15] is a single-stage network model which estimates classification score and bounding boxes straight from input image with the help a simple CNN. The model divides the input image into fixed number of grids. Each grid cell predicts fixed number of bounding boxes with a confidence score. The confidence score is calculated by multiplying the object detection probability with the IoU ratio of predicted bounding box and the ground-truth boxes. If a bounding box got the class probability beyond a threshold value, then the box is selected and used further for object localization (Fig. 8).

Training details: YOLO has 24 conv layers and two FC layers. The authors have increased the resolution of input image and used batch normalization [42] to the bounding box coordinates, box height and box width to increase detection accuracy. The final layer of the network used linear activation function and rest of the layers used leaky rectified linear activation [43]. Uses of sum squared error in the output layer made their model to be optimized easily. Though YOLO predicts multiple bounding boxes per grid cell, the authors have chosen one object prediction per bounding box during training. These strategies lead to better prediction with respect to sizes, aspect ratio, classes and overall recall. They have trained/validated/tested their network on PASCAL VOC 2007 and 2012 datasets. For training, they have used a mini-batch size of 64, a momentum of 0.9 and a weight decay of 0.005. Their initial learning rate was 0.001, then gradually increased it to 0.01 for 70 epochs, then decreased it by a factor of 10 for each 30 epochs. To avoid overfitting, they have used dropout [44] and heavy data augmentation.

SSD: The single-shot multibox detection (SSD) [45] network takes an entire image as input and moves it across multiple conv layers with different-sized filters (10×10, 5×5 and 3×3) as shown in Fig. 9. To predict the bounding boxes, convolutional

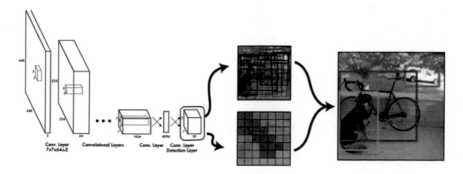

Fig. 8 Architecture of YOLO [15]

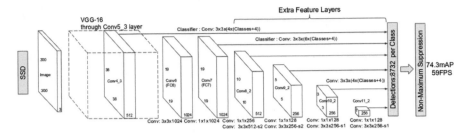

Fig. 9 Architecture of SSD [45]

feature maps of different levels of the network are used . At individual level, the feature maps are convolved using an extra feature layer of 3×3 filter to produce bounding boxes. Like the anchor boxes (default box) of the Fast R-CNN, anchor box of SSD has four parameters: two parameters for the coordinates of the center, one for the width and one for the height. At the time of predicting bounding boxes, the model produces a vector of class probabilities of objects. To incorporate multi-scale object detection, SSD predicts bounding boxes from multiple convolutional layers. This object detector achieves a good balance between speed and accuracy.

Training details: During training, for each ground-truth box, the authors have selected those anchor boxes that vary with respect to location, aspect ratio and scales. Then, they have matched anchor with ground-truth box with the best Jaccard overlap [46]. They have selected those anchor boxes whose Jaccard overlap to the ground truth is >0.5. To handle different object scales, they have used different feature maps from different layers of their network. They have designed the anchors in such a way that certain feature maps learn to be responsive to particular scales of the objects. They have trained their model using hard negative mining with negative and positive default box ratio of 3:1 and using heavy data augmentation.

YOLO9000: YOLO9000 [47] is a modified version of YOLO which is able to detect more than 9000 object categories in real-time by jointly optimizing detection and classification. The authors have used WordTree [48] hierarchy to combine data from various sources such as ImageNet and MS COCO [36] dataset and proposed an algorithm to jointly train the network for optimization on those different datasets simultaneously. The authors have also made some improvements on the previous YOLO model to enhance its performances keeping its original speed.

Training details: The authors first produced YOLOv2 [49] improving the basic YOLO system. Then, they have trained their model using joint training algorithm on more than 9000 classes from ImageNet and from MS COCO detection data. They have used batch normalization instead of dropout, and their input image size was 416×416 to capture center part of an image. They have removed a pooling layer from the network to capture high-resolution features. Anchor boxes are used for predicting bounding boxes. The dimensions of anchor box are chosen using k-means clustering on the training set. They have used fine-grained features to detect smaller object and also

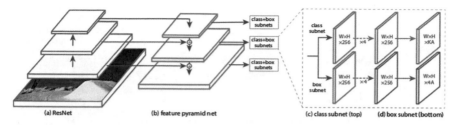

| (a) ResNet | (b) feature pyramid net | (c) class subnet (top) | (d) box subnet (bottom) |

Fig. 10 Architecture of RetinaNet [16]

used multi-scale training by choosing new image size randomly per 10 batches to make the model scale invariant.

RetinaNet: RetinaNet [16] is a simple one-stage, unified object detector which works on dense sampling of object locations in an input image. As shown in Fig. 10, the model consists of a backbone network and two task-specific sub-networks. As backbone network, they have used feature pyramid network (FPN) [35]. On the backbones output, the first sub-network performs convolutional object classification and the second sub-network performs convolutional bounding box regression. These sub-networks are basically small fully convolutional network (FCN) [34] attached to each FPN level for parameter sharing.

Training details: The authors have proposed a loss function called focal loss to handle one-stage object detection scenario where the network suffers from extreme foreground and background class imbalance during training. They have used this loss function on the output of the classification subnet. RetinaNet is trained with synchronized SGD over 8 GPUs with mini-batch of size 16 (two images per mini-batch per GPU). The model training is initiated with learning rate of 0.01 for first 90k iterations. Then, the learning is decreased by 1/10 for next 60k iteration and again decreased at the previous rate for rest 80k iterations. A momentum of 0.9 and a weight decay of 0.0001 are used. The training loss of their model is the aggregation of the focal loss for classification and the standard smooth L1 loss for box regression.

RefineDet: RefineDet [50], as shown in Fig. 11 is a CNN-based single-shot object detection model. This network consists two interconnected modules—(i) the anchor refinement module (ARM) and (ii) the object detection module (ODM). First, the ARM module reduces search space for the classifier by filtering negative anchors and roughly adjusts the sizes and locations of the anchors for providing better initialization for the successive regressors. Taking the refined anchors from the previous module as input, the ODM module improves the regression accuracy and predicts multiple class labels. The transfer connected block (TCB) transfers the features of the ARM module to predict class labels, sizes and locations of objects in the ODM module.

Training details: The authors have used VGG-16 and ResNet 101 [51] network as base network. These networks are pre-trained on ImageNet classification and

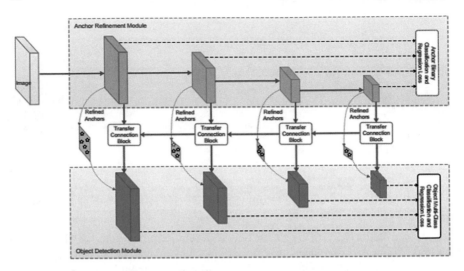

Fig. 11 Architecture of RefineDet [50]

detection dataset. Instead of sixth and seventh FC layers of VGG-16, they have used conv layers. Also to capture high-level information and detecting objects in multiple scales, they have added two conv layers at the end of the VGG-16 network and one residual block at the end of the ResNet-101. The parameters of extra two layers of VGG-16 are initialized using "xavier" [52] method. To deal with different object scales, they have selected four feature layers with stride size 8, 16, 32 and 64 pixels for both CNN networks. Then, they have associated those strides with several different anchors for prediction. The authors determined the matching of anchor and ground-truth box using Jaccard overlap following [45]. Also they have trained their model using hard negative mining following [45]. They set the default training mini-batch size to 32 and then fine-tuned the network using SGD with 0.9 momentum and 0.0005 weight decay.

3 Comparative Result

As we have reviewed object detection models of two different types (two stage and one stage), performance comparison of those models is tough. In real-life application, performance basically refers to accuracy and speed. We have seen here that performance of a object detection model depends on different aspects such as feature extractor network, input image resolution, matching strategy and IoU threshold, non-maximum suppression, the number of proposals or predictions, boundary box encoding, loss function, optimization function, chosen value for training hyperparameters. In Table 1, we have shown comparative performance of different object detection models on PASCAL VOC and MS COCO datasets. From second to sixth

Table 1 Comparison of different object detection models: column 1 shows the name of the object detection model, columns 2–6 show object detection accuracy (mAP) on a different dataset

Object Detection Model	Pascal VOC 2007 (%)	Pascal VOC 2010 (%)	Pasca VOC 2012 (%)	MS COCO 2015 (%)	MS COCO 2016 (%)	Real time detection	Number of stages
R-CNN	58.5	53.7	53.3			No	Two
SPP-net	59.2					No	Two
Fast R-CNN	70.0	68.8	68.4			No	Two
Faster R-CNN	73.2		70.4			No	Two
FPN					35.8	No	Two
Mask R-CNN					43.4	No	Two
YOLO	63.4		57.9			Yes	One
SSD	81.6		80.0	46.5		No	One
YOLO9000	78.6		73.4	21.6		Yes	One
RetinaNet					40.8	Yes	One
RefineDet	85.6		86.8		41.8	Yes	One

columns of the table represent the mean average precision (mAP) in measuring accuracy of object detection. Last two columns give an overview of those model's speed and organizational approach.

4 Conclusion

In this chapter, we have explored the advancements of different object detection models based on CNN. Among various state-of-the-art models, we have presented major architectural improvements of some well-accepted models. In our review, we have categorized those models according to two different approaches: two-stage approach and one-stage approach. In terms of gaining accuracy in object detection, two-stage models left behind the one-stage models. On the other hand, one-stage models are faster than two-stage models in terms of inference time. External region proposal network of R-CNN, SPP-net and Fast R-CNN made them slower. Use of CNN-based region proposal network helped Faster R-CNN to overcome that problem. Mask R-CNN has incorporated instance segmentation along with object detection. YOLO, SSD showed us a way of fast as well as robust object detection. RetinaNet has focused on improvement in loss function to gain higher accuracy. RefineDet used a combination of the benefits of both two-stage and one-stage approaches, and it has achieved state-of-the-art performance. Through this chapter, we have explored that the advancements of various state-of-the-art models mainly depend on architectural development of CNN models, newer detection architecture, different approaches for pooling, different optimization algorithm and model training approaches, novel loss function, etc. The continuous improvements of different object detection models give us the hope toward better object detection model with higher accuracy, more robustness and faster as well as real-time object detection.

References

1. Lowe, D. G. (2004). Distinctive image features from scale-invariant keypoints. *International Journal of Computer Vision, 60*(2), 91–110. https://doi.org/10.1023/B:VISI.0000029664. 99615.94.
2. Dalal, N., & Triggs, B. (2005). Histograms of oriented gradients for human detection. In *2005 IEEE Computer Society Conference on Computer Vision and Pattern Recognition (CVPR'05)* (Vol. 1, pp. 886–893).
3. Everingham, M., Van Gool, L., Williams, C. K. I., Winn, J., & Zisserman, A. (2010). The pascal visual object classes (VOC) challenge. *International Journal of Computer Vision, 88*(2), 303–338. https://doi.org/10.1007/s11263-009-0275-4.
4. Fukushima, K. (1980). Neocognitron: A self-organizing neural network model for a mechanism of pattern recognition unaffected by shift in position. *Biological Cybernetics, 36*(4), 193–202. https://doi.org/10.1007/BF00344251.
5. LeCun, Y., Boser, B., Denker, J. S., Henderson, D., Howard, R. E., Hubbard, W., et al. (1989). Backpropagation applied to handwritten zip code recognition. *Neural Computation, 1*(4), 541–551. https://doi.org/10.1162/neco.1989.1.4.541.
6. Lecun, Y., Bottou, L., Bengio, Y., & Haffner, P. (1998). Gradient-based learning applied to document recognition. In *Proceedings of the IEEE* (pp. 2278–2324).
7. Bottou, L. (2010). Large-scale machine learning with stochastic gradient descent. In *Proceedings of COMPSTAT'2010* (pp. 177–186). Heidelberg: Physica-Verlag HD.
8. Cun, Y. L. (1988). A theoretical framework for back-propagation.
9. Ghosh, A., Sufian, A., Sultana, F., Chakrabarti, A., & De D. (2020). Fundamental concepts of convolutional neural network. In *Recent trends and advances in artificial intelligence and Internet of Things* (pp. 519–567).
10. Krizhevsky, A., Sutskever, I., & Hinton, G. E. (2012). Imagenet classification with deep convolutional neural networks. In *Advances in neural information processing systems 25* (pp. 1097–1105).
11. Sultana, F., Sufian, A., & Dutta, P. (2018). Advancements in image classification using convolutional neural network. In *2018 Fourth International Conference on Research in Computational Intelligence and Communication Networks (ICRCICN)* (pp. 122–129), Nov 2018.
12. Girshick, R. B., Donahue, J., Darrell, I., & Malik, J. (2014). Rich feature hierarchies for accurate object detection and semantic segmentation. In *2014 IEEE Conference on Computer Vision and Pattern Recognition* (pp. 580–587).
13. He, K., Zhang, X., Ren, S., & Sun, J. (2014). Spatial pyramid pooling in deep convolutional networks for visual recognition. *CoRR.* arxiv.org/abs/1504.08083.
14. Girshick, R. B. (2015). Fast R-CNN. *CoRR.* arxiv.org/abs/1504.08083.
15. Redmon, J., Divvala, S. K., Girshick, R. B., & Farhadi, A. (2015). You only look once: Unified, real-time object detection. *CoRR.* arxiv.org/abs/1506.02640.
16. Lin, T., Goyal, P., Girshick, R. B., He, K., & Dollár, P. (2017). Focal loss for dense object detection. *CoRR.* arxiv.org/abs/1708.02002.
17. Wu, B., Iandola, F. N., Jin, P. H., & Keutzer, K. (2017). Squeezedet: Unified, small, low power fully convolutional neural networks for real-time object detection for autonomous driving. *CoRR.* arxiv.org/abs/1612.01051.
18. Woo, S., Hwang, S., & Kweon, I. S. (2017). Stairnet: Top-down semantic aggregation for accurate one shot detection. *CoRR.* arxiv.org/abs/1709.05788.

19. Li, W., Liu, K., Yan, L., Cheng, F., Lv, Y., & Zhang, L. (2019). FRD-CNN: Object detection based on small-scale convolutional neural networks and feature reuse. *Scientific Reports*, *9*(1), 1–12.
20. Uijlings, J., van de Sande, K., Gevers, I., & Smeulders, A. (2013). Selective search for object recognition. *International Journal of Computer Vision*. http://www.huppelen.nl/publications/selectiveSearchDraft.pdf.
21. Wolberg, G. (1994). *Digital image warping* (1st ed.). Los Alamitos, CA, USA: IEEE Computer Society Press.
22. Hearst, M. A., Dumais, S. T., Osuna, E., Platt, J., & Scholkopf, B. (1998). Support vector machines. *IEEE Intelligent Systems and their Applications*, *13*(4), 18–28.
23. Deng, J., Dong, W., Socher, R., Li, L.-J., Li, K., Fei-Fei, L. (2009). ImageNet: A large-scale hierarchical image database. In *CVPR09*.
24. Russakovsky, O., Deng, J., Su, H., Krause, J., Satheesh, S., Ma, S., et al. (2015). Imagenet large scale visual recognition challenge. *IJCV*, *115*(3), 211–252. https://doi.org/10.1007/s11263-015-0816-y.
25. Bucher, M., Herbin, S., & Jurie, F. (2016). Hard negative mining for metric learning based zero-shot classification. *CoRR*. arxiv.org/abs/1608.07441.
26. Grauman, K., & Darrell, T. (2005). The pyramid match kernel: discriminative classification with sets of image features. In *Tenth IEEE International Conference on Computer Vision (ICCV'05) Volume 1* (Vol. 2, pp. 1458–1465).
27. Lazebnik, S., Schmid, C., & Ponce, J. (2006). Beyond bags of features: Spatial pyramid matching for recognizing natural scene categories. In *IEEE Conference on Computer Vision & Pattern Recognition (CPRV '06)* (pp. 2169 – 2178). New York, United States: IEEE Computer Society, June 2006. https://hal.inria.fr/inria-00548585.
28. van de Sande, K. E. A., Uijlings, J. R. R., Gevers, T., & Smeulders, A. W. M. (2011). Segmentation as selective search for object recognition. In *2011 International Conference on Computer Vision* (pp. 1879–1886), November 2011.
29. Felzenszwalb, P. F., Girshick, R. B., McAllester, D., & Ramanan, D. (2010). Object detection with discriminatively trained part-based models. *IEEE Transactions on Pattern Analysis and Machine Intelligence*, *32*(9), 1627–1645.
30. Simonyan, K., & Zisserman A. (2014). Very deep convolutional networks for large-scale image recognition. *CoRR*. arxiv.org/abs/1409.1556.
31. Chatfield, K., Simonyan, K., Vedaldi, A., & Zisserman, A. (2014). Return of the devil in the details: Delving deep into convolutional nets. *CoRR*. arxiv.org/abs/1405.3531.
32. Ren, S., He, K., Girshick, R. B., Zhang, X., & Sun, J. (2015). Object detection networks on convolutional feature maps. *CoRR*. arxiv.org/abs/1504.06066.
33. Long, J., Shelhamer, E., & Darrell, T. (2014). Fully convolutional networks for semantic segmentation. *CoRR*. arxiv.org/abs/1411.4038.
34. Shelhamer, E., Long, J., & Darrell, T. (2017). Fully convolutional networks for semantic segmentation. *IEEE Transactions on Pattern Analysis and Machine Intelligence*, *39*(4), 640–651. https://doi.org/10.1109/TPAMI.2016.2572683.
35. Lin, T., Dollár, P., Girshick, R. B., He, K., Hariharan, B., & Belongie, S. J. (2016). Feature pyramid networks for object detection. *CoRR*. arxiv.org/abs/1612.03144.
36. Lin, T.-Y., Maire, M., Belongie, S., Hays, J., Perona, P., Ramanan, D., et al. (2014). Microsoft COCO: Common objects in context. In D. Fleet, T. Pajdla, B. Schiele, & T. Tuytelaars (Eds.), *Computer vision—ECCV 2014* (pp. 740–755). Cham: Springer.
37. Iglovikov, V. I., Seferbekov, S. S., Buslaev, A. V., & Shvets, A. (2018). Ternausnetv2: Fully convolutional network for instance segmentation. *CoRR*. arxiv.org/abs/1806.00844.
38. Chen, L., Hermans, A., Papandreou, G., Schroff, E., Wang, P., & Adam, H. (2017). Masklab: Instance segmentation by refining object detection with semantic and direction features. *CoRR*. arxiv.org/abs/1712.04837.
39. He, K., Gkioxari, G., Dollár, P., & Girshick, R. B.: Mask R-CNN. In *2017 IEEE International Conference on Computer Vision (ICCV)* (pp. 2980–2988).

40. He, K., Zhang, X., Ren, S., & Sun, J. (2015). Deep residual learning for image recognition. arxiv.org/abs/1512.03385.
41. Xie, S., Girshick, R. B., Dollár, P., Tu, Z., & He, K. (2016). Aggregated residual transformations for deep neural networks. *CoRR*. arxiv.org/abs/1611.05431.
42. Bjorck, J., Gomes, C. P., & Selman, B. (2018). Understanding batch normalization. *CoRR*. arxiv.org/abs/1806.02375.
43. Xu, B., Wang, N., Chen, T., & Li, M. (2015). Empirical evaluation of rectified activations in convolutional network. *CoRR*. arxiv.org/abs/1505.00853.
44. Srivastava, N., Hinton, G. E., Krizhevsky, A., Sutskever, I., Salakhutdinov, R. (2014). Dropout: A simple way to prevent neural networks from overfitting. *Journal of Machine Learning Research, 15*(1), 1929–1958. http://www.cs.toronto.edu/~rsalakhu/papers/srivastava14a.pdf.
45. Liu, W., Anguelov, D., Erhan, D., Szegedy, C., Reed, S. E., Fu, C.-Y., & Berg, A. C. (2016). SSD: Single shot multibox detector. In *ECCV*.
46. Erhan, D., Szegedy, C., Toshev, A., & Anguelov, D. (2013). Scalable object detection using deep neural networks. *CoRR*. arxiv.org/abs/1312.2249.
47. Redmon J., & Farhadi, A. (2017). Yolo9000: Better, faster, stronger. In *2017 IEEE Conference on Computer Vision and Pattern Recognition (CVPR)* (pp. 6517–6525).
48. Wattenberg, M., & Viégas, F. B. (2008). The word tree, an interactive visual concordance. *IEEE Transactions on Visualization and Computer Graphics, 14*.
49. Nakahara, H., Yonekawa, H., Fujii, T., & Sato, S. (2018). A lightweight YOLOv2: A binarized CNN with a parallel support vector regression for an FPGA. In *Proceedings of the 2018 ACM/SIGDA International Symposium on Field-Programmable Gate Arrays, Series FPGA'18* (pp. 31–40). New York, NY, USA: ACM. https://doi.org/10.1145/3174243.3174266.
50. Zhang, S., Wen, L., Bian, X., Lei, Z., & Li, S. Z. (2017). Single-shot refinement neural network for object detection. *CoRR*. arxiv.org/abs/1711.06897.
51. He, K., Zhang, X., Ren, S., & Sun, J. (2016). Deep residual learning for image recognition. In *The IEEE Conference on Computer Vision and Pattern Recognition (CVPR)*, June 2016.
52. Glorot, X., & Bengio, Y. (2010). Understanding the difficulty of training deep feedforward neural networks. ' In *Proceedings of the Thirteenth International Conference on Artificial Intelligence and Statistics, Series Proceedings of Machine Learning Research PMLR* (Vol. 9, pp. 249–256).

Cancerous Cell Detection Using Affine Transformation with Higher Accuracy and Sensitivity

Soumen Santra, Joy Bhattacherjee, and Arpan Deyasi

Abstract Cancerous cells are detected by affine transformation by working on multi-resolution, multi-lateral, multi-cluster image patterns for prediction analysis. Accuracy, sensitivity and specificity of a few images are calculated for the affected cells, where results provide less confusion value of transformed images than the originals. Authenticity of the cell is investigated by computing true and false positive and negative rates. Difference in maximum peak amplitude for Z-transformed image and Z-transformed affine image is calculated to evaluate the importance of the transform procedure. From the assessment of image, status of the cancerous cell can be easily revealed, whether it is benign or malignant, and the corresponding decision can be taken for shielding of other cells.

Keywords Affine transformation · Cancerous cell · Z-transform · Assessment parameters · Accuracy · Sensitivity · Specificity

1 Introduction

The study of medical image classification relates with various aspects, one of the aspects among them is dimension based features description technique [1]. These regions based multi-dimensional multi-lateral descriptor acts as morph-metric tool which also can explain contour analysis for close convex images [2]. So we can analyse a medical image through their edges or segments or features [3, 4]. Several

S. Santra (✉)
Department of Computer Application, Techno International New Town, Kolkata 700156, India
e-mail: soumen70@gmail.com

J. Bhattacherjee
Department of Electronic Science, A.P.C College, Barasat 700131, India
e-mail: joybhatta1000@gmail.com

A. Deyasi
Department of Electronics and Communication Engineering, RCC Institute of Information Technology, Kolkata 700015, India
e-mail: deyasi_arpan@yahoo.co.in

© Springer Nature Singapore Pte Ltd. 2020
J. K. Mandal and S. Banerjee (eds.), *Intelligent Computing: Image Processing Based Applications*, Advances in Intelligent Systems and Computing 1157,
https://doi.org/10.1007/978-981-15-4288-6_2

17

statistical techniques are incorporated for edge detection purpose [2]. Here, we give a brief knowledge regarding various variants and then invariant transformations such as affine, 'Z' which are working on multi-resolution, multi-lateral, multi-cluster image patterns. Sometimes, we get same information from extrinsic structural components [5] of different clusters as well as different information from several intrinsic structural components of same cluster regions [6]. So we need invariant descriptor tools which work upon locally and globally same throughout the image pattern. This tool gives a prediction-based idea for early detection of cell-related diseases, and those are now challenging issues in medical image technology [7, 8]. All the outputs are stored as parameters upon which we can make the prediction. So the analysis of data plays a big role by which prediction happens in an easy way. This tool forecasts us to tell the scenarios or fact-finding data by which we can predict the future stages of cells of medical image output. Preparation of manuscripts which are to be reproduced by photo-offset requires special care. Papers submitted in a technically unsuitable form will be returned for retyping or cancelled if the volume cannot otherwise be finished on time.

2   Affine Transformation

Works on affine transform are initiated more than a decade ago when Flusser and his co-workers [9] calculated several invariants under affine transformation from second- and third-order moments and are used for pattern recognition. Novel interpolation technique is proposed for recognition and reconstruction of images [10] where workers claimed that the method provides highest throughput with minimum complexity. The technique is suitable for graphics manipulation. For the same purpose, two-dimensional cell animation technique is also utilized [11], though it involves higher time consumption and lower throughput. In subsequent time, researchers train neural network for affine transformed predicted parameters learning [12] using singular-value decomposition technique and interval arithmetic where sampling of space is performed for the affined transformed views. New affine transformation method is also proposed for image coding [13] with the claim of improved PSNR and CR. Fast affine transformation method is proposed [14] where computational complexity is reduced by removing the number of multiplications, instead by considering the co-relation between the relations of neighbouring pixels and therefore can be implemented for real-time applications. A few experimental results related with different mathematical procedures are reported [15] which are later applied in 3D biomedical images [16] through replication of the voxel location and therefore greatly reduces the number of matrix multiplication, along with increment of speed. Affine invariant functions are also used in spatial domain by eliminating the point-of-choice effect inside the contour [17], which can effectively be used for object classification. Concept of pipelined architecture is used [18] to reduce processing time which is later verified by hardware set-up and therefore can be applied in optical quadrature microscopy. The drawback of moment transformation method in terms of information redundancy is removed by workers by proposing combined invariants of orthogonal Legendre moments [19] where improved robustness is achieved. For large planar antenna array, modified alternating projection method is applied for faster synthesization [20] which results in less number of devices to produce same accuracy. Very recently, real-time image registration problem is tackled by continuous piecewise affine transformation [21], and new hardware architecture is made to reduce time of ASIFT algorithm [22] which also simultaneously decreases memory space for image filtering by adopting skewed kernel technique. In this year, the method is applied to wireless sensor network [23, 24] for wider coverage area through a novel optimization technique [23, 24] by implementing various mutation strategies to different sub-populations.

3 Objective

In the present chapter, application of affine transform in the case of sensitive medical imaging is proposed and is compared to the original image and affine transformed image using some image assessment parameters and Z-transform. It is observed that the Z-transform of affine transformed image becomes more useful to detect the affected cell location with an immense value difference. Affine transform helps us to fix the dimension of an image by geometric rotation. In real time, MRI scan is made with the help of a probe camera from inside of human body. This technique may be biased due to many circumstances and also the image can change its geometric dimension, and it is not possible to take another live scan on that affected area again.

4 Mathematical Transformations

Affine transform is basically a 2D geometric transformation which creates alienation between pixel intensity values of input and output images. For that purpose, it applies a linear combination of translation, rotation, scaling or shearing operations [25]. It is a linear mapping method that preserves points, straight lines and planes. If we consider a set of parallel lines undergoes affine, it will remain parallel after the operation. It is more generally used to modify the distortions obtained geometrically or the deformations generated owing to non-ideal camera positions.

In common, the affine transformation 'A' in three-dimensional function domain is defined by the following linear equations:

$$x' = ax + by + cz + d$$
$$y' = ex + fy + gz + h$$
$$z' = kx + ly + mz + n$$

In matrix notation, we can write

$$A \begin{bmatrix} x \\ y \\ z \end{bmatrix} = \begin{bmatrix} x' \\ y' \\ z' \end{bmatrix} = \begin{bmatrix} a & b & c \\ e & f & g \\ k & l & m \end{bmatrix} \begin{bmatrix} x' \\ y' \\ z' \end{bmatrix} + \begin{bmatrix} d \\ h \\ n \end{bmatrix}$$

This shows that the images are related by twelve parameters, nine of which resembles the linear transformations like scaling, rotation, shears and the other three represents translation in two orthogonal directions. The leading part of the right-hand side of the above equation represents the linear part of the transformation, and another part represents the translational part of the same.

4.1 Special Cases in Affine Transform

4.1.1 Scaling/Zooming

$$\text{If } \begin{bmatrix} a & b & c \\ e & f & g \\ k & l & m \end{bmatrix} = \begin{bmatrix} p_x & 0 & 0 \\ 0 & p_y & 0 \\ 0 & 0 & p_z \end{bmatrix} \text{ and } \begin{bmatrix} d \\ h \\ n \end{bmatrix} = \begin{bmatrix} 0 \\ 0 \\ 0 \end{bmatrix}$$

Then,

$$x' = p_x x$$
$$y' = p_y y$$
$$z' = p_z z$$

p_x, p_y, p_z have some arithmetical values to scale or zoom the image.

4.1.2 Rotation

$$\text{If } \begin{bmatrix} a & b & c \\ e & f & g \\ k & l & m \end{bmatrix} = \begin{bmatrix} \cos\theta & -\sin\theta & 0 \\ \sin\theta & \cos\theta & 0 \\ 0 & 0 & 1 \end{bmatrix} \text{ and } \begin{bmatrix} d \\ h \\ n \end{bmatrix} = \begin{bmatrix} 0 \\ 0 \\ 0 \end{bmatrix}$$

Then,

$$x' = x\cos\theta - y\sin\theta$$
$$y' = x\sin\theta + y\cos\theta$$
$$z' = z$$

4.1.3 Shear (Vertical)

If
$$\begin{bmatrix} a & b & c \\ e & f & g \\ k & l & m \end{bmatrix} = \begin{bmatrix} 1 & S_v & 0 \\ 0 & 1 & 0 \\ 0 & 0 & 1 \end{bmatrix} \quad \text{and} \quad \begin{bmatrix} d \\ h \\ n \end{bmatrix} = \begin{bmatrix} 0 \\ 0 \\ 0 \end{bmatrix}$$

Then,

$$x' = x + S_v y$$
$$y' = y$$
$$z' = z$$

4.1.4 Shear (Horizontal)

If
$$\begin{bmatrix} a & b & c \\ e & f & g \\ k & l & m \end{bmatrix} = \begin{bmatrix} 1 & 0 & 0 \\ S_h & 1 & 0 \\ 0 & 0 & 1 \end{bmatrix} \quad \text{and} \quad \begin{bmatrix} d \\ h \\ n \end{bmatrix} = \begin{bmatrix} 0 \\ 0 \\ 0 \end{bmatrix}$$

Then,

$$x' = x$$
$$y' = S_h x + y$$
$$z' = z$$

4.1.5 Shear (Horizontal and Vertical Simultaneously)

$$\text{If} \begin{bmatrix} a & b & c \\ e & f & g \\ k & l & m \end{bmatrix} = \begin{bmatrix} 1 & S_v & 0 \\ S_h & 1 & 0 \\ 0 & 0 & 1 \end{bmatrix} \text{ and } \begin{bmatrix} d \\ h \\ n \end{bmatrix} = \begin{bmatrix} 0 \\ 0 \\ 0 \end{bmatrix}$$

Then,

$$x' = x + S_v y$$
$$y' = S_h x + y$$
$$z' = z$$

4.2 Z-Transform

It is a transformation where it converts discrete-time signal into frequency time-domain representation. We know a signal always is a combination of real and complex components. In this transform, we can divide both the parts individual on the basis of bandwidth. It is quite equivalent to Laplace and Fourier transform, but if we plot each bandwidth on the basis of the total length of the bandwidth, then we can easily identify each part of the signal. These properties are also same in case of image. It is one of the forms of standardization for scaling the image signal. This Z score defines the standard deviation of each bandwidth for image components against its size where we can identify real and imaginary parts of that image signal. This z-plane and Z-domain helps to extract more extra features by which we can get more information about the image properties. Z-transform gives the combination of real part of lower frequency (LF), real part of horizontal frequency (HF), real part of vertical frequency (VF) and complex part of horizontal frequency (HF). It also gives combination of all real parts and combination of all imaginary part with negation of imaginary part of HF.

5 Flow Diagram and Algorithm

Total process flow for the cell detection is described below with the diagram

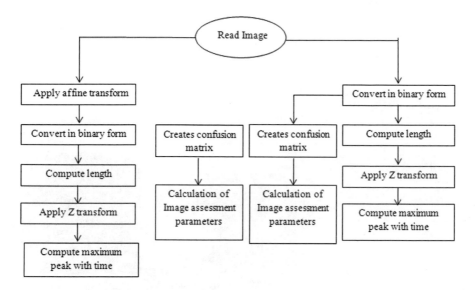

Based on these steps, the following algorithm is developed:

1. Read the image and image address stored in variable 'A'.
2. Image stored in variable 'A' is displayed.
3. Apply '2D affine transform' of 'A' with respect to different matrices (e.g. scaling, rotation) and stored in 'B'.
4. Image stored in variable 'B' is displayed.
5. 'A' is converted into binary form and stored in 'bw1'.
6. 'bw1' is stored in a $m1 \times n1$ matrix [$m1=$ size of row, $n1=$ size of column].
7. Find Z score () of 'bw1' and stored in 'z1'.
8. Length of 'z1' stored in 'p1'.
9. Zscore is plotted (length of 'z1' vs. 'z1').
10. Confusion matrix is created of 'bw1' and calculated the image assessment parameters.
11. 'B' is binaries and stored in 'bw2'.
12. 'bw2' is stored in a $m2 \times n2$ matrix [$m2=$ size of row, $n2=$ size of column].
13. Find Z score () of 'bw2' and stored in 'z2'.
14. Length of 'z2' stored in 'p2'.
15. Z score is plotted (length of 'z2' vs. 'z2').
16. Confusion matrix is created of 'bw2' and calculated the image assessment parameters.

6 Results

In this section, we consider ten images and their various affine transform with respect to transformation matrix based on MRI of brain tumour, mammogram and other scanned images of various parts from human body.

In Table 1, we show various transformation matrices of affine transformation of sensitive medical image.

Table 1 Comparative study of different affine transform with respect to transformation matrix of sensitive medical images

Cases and transformation matrix format	Original image	Affine transformed image
1. Scaling $$\begin{bmatrix} 2.5 & 0 & 0 \\ 0 & 2.3 & 0 \\ 0 & 0 & 1 \end{bmatrix}$$	(Breast cancer mammogram image) (By Howard.r.prtty—Own work, CC BY 4.0, https://commons.wikimedia.org/w/index.php?curid=64375620)	
2. Rotation $$\begin{bmatrix} \cos d(50) & -\sin d(50) & 0 \\ \sin d(50) & \cos d(50) & 0 \\ 0 & 0 & 1 \end{bmatrix}$$	(Brain tumour MRI) (By Novaksean—Own work, CC BY-SA 4.0, https://commons.wikimedia.org/w/index.php?curid=44348140)	
3. Shear (Vertical) $$\begin{bmatrix} 1 & 1 & 0 \\ 0 & 1 & 0 \\ 0 & 0 & 1 \end{bmatrix}$$	(Brain tumour MRI) (By © Nevit Dilmen, CC BY-SA 3.0, https://commons.wikimedia.org/w/index.php?curid=18741460)	

(continued)

Table 1 (continued)

Cases and transformation matrix format	Original image	Affine transformed image
4. Shear (Horizontal) $$\begin{bmatrix} 1 & 0 & 0 \\ 1 & 1 & 0 \\ 0 & 0 & 1 \end{bmatrix}$$	(Brain tumour MRI) (By © Nevit Dilmen, CC BY-SA 3.0, https://commons.wikimedia.org/w/index.php?curid=18633668)	
5. Shear (Vertical and horizontal) $$\begin{bmatrix} 1 & 0.5 & 0 \\ 0.5 & 1 & 0 \\ 0 & 0 & 1 \end{bmatrix}$$	(Breast cancer mammogram image) (By Unknown photographer—This image was released by the National Cancer Institute, an agency part of the National Institutes of Health, with the ID 2703 (image) (next)., Public Domain, https://commons.wikimedia.org/w/index.php?curid=24052431)	
6. Shear (Vertical) $$\begin{bmatrix} 2.5 & 1 & 0 \\ 0 & 1 & 0 \\ 0 & 0 & 1 \end{bmatrix}$$	(Breast cancer mammogram image) (By Unknown photographer—This image was released by the National Cancer Institute, an agency part of the National Institutes of Health, with the ID 2510 (image) (next)., Public Domain, https://commons.wikimedia.org/w/index.php?curid=24052516)	

(continued)

Table 1 (continued)

Cases and transformation matrix format	Original image	Affine transformed image
7. Rotation $$\begin{bmatrix} \cos d(60) & -\sin d(60) & 0 \\ \sin d(60) & \cos d(60) & 0 \\ 0 & 0 & 1 \end{bmatrix}$$	(Brain tumour MRI) (By © Nevit Dilmen, CC BY-SA 3.0, https://commons.wikimedia. org/w/index.php?curid= 18633690)	
8. Scaling/Zooming $$\begin{bmatrix} 2.5 & 0 & 0 \\ 0 & 2.3 & 0 \\ 0 & 0 & 1 \end{bmatrix}$$	(Brain tumour MRI) (By Hellerhoff—Own work, CC BY-SA 4.0, https://commons. wikimedia.org/w/index.php? curid=35028568)	
9. Shear (Vertical and horizontal) $$\begin{bmatrix} 1 & 0.5 & 0 \\ 0.5 & 1 & 0 \\ 0 & 0 & 1 \end{bmatrix}$$	(Brain tumour MRI) (By © Nevit Dilmen, CC BY-SA 3.0, https://commons.wikimedia. org/w/index.php?curid= 18477907)	

(continued)

Table 1 (continued)

Cases and transformation matrix format	Original image	Affine transformed image
10. Shear (Horizontal) $$\begin{bmatrix} 1 & 0 & 0 \\ 2.8 & 1 & 0 \\ 0 & 0 & 1 \end{bmatrix}$$	(Breast cancer mammogram image) (By Unknown photographer—This image was released by the National Cancer Institute, an agency part of the National Institutes of Health, with the ID 2706 (image) (next)., Public Domain, https:// commons.wikimedia.org/w/ index.php?curid=24052396)	

The first image is a mammogram scan of breast cancer cell, which is scaled 2.5 and 2.3 times on the vertical and horizontal axis (*x-axis* and *y*-axis) and causes to zoom the image. The second image is MRI scan of brain tumour which is rotated 50° from its initial position. The third image is also an MRI scan of brain tumour which is sheared vertically 1 unit. The fourth image is an MRI scan of a brain tumour cell which is sheared horizontally 1 unit. The fifth image is a breast cancer image which is sheared both vertically and horizontally 0.5 units and so on. All the transformation is done using preferred transformation matrix (Table 2).

Therefore, we show here how the position of cancerous cell is becoming easier to find out in Z-transform of affine transformed image relatively than the original image. It will be also evident that the number of maximum peak of amplitude (which denotes the abrupt change in pixel intensity, i.e. location of cancerous cell) becomes less than original image with unadulterated location in the case of Z-transform of affine transformed image, which helps to find out the location of cancerous cell with less confusion value. The amplitude of maximum peaks is also increased in the case of Z-transformed image obtained from affine transformed image; this indicates that the change in image assessment parameters like true positive rate, true negative rate helps us to detect the located cell seriously affected or just a normal cell located in that position. This occurs as the location of cancerous cell is similarly geometrically transformed with respect to affine transform.

Table 2 Comparative study of Z-transforms obtained from original image and affine transformed image

Original image	Z-transform of original image	Z-transform of affine transformed image
(Brain tumour MRI) (By Novaksean—Own work, CC BY-SA 4.0, https://commons. wikimedia.org/w/index. php?curid=44348140)	 Maximum peak amp.: 21.142 units (4 times) Time: 24–37 s	 Maximum peak amp.: 24.440 units (1 time) Time: 24 s
(Brain tumour MRI) (By © Nevit Dilmen, CCBY-SA 3.0, https:// commons.wikimedia.org/ w/index.php?curid= 18741460)	 Maximum peak amp.: 15.811 units (5 times) Time: 448–919 s	 Maximum peak amp.: 54.037 units (1 time) Time: 905 s
(Brain tumour MRI) (By © Nevit Dilmen, CC BY-SA 3.0, https:// commons.wikimedia.org/ w/index.php?curid= 18477907)	 Maximum peak amp.: 16.417 units (12 times) Time: 448–578 s	 Maximum peak amp.: 15.811 units (1 time) Time: 448 s

(continued)

Table 2 (continued)

Original image	Z-transform of original image	Z-transform of affine transformed image
(Brain tumour MRI) (By © Nevit Dilmen, CC BY-SA 3.0, https:// commons.wikimedia.org/ w/index.php?curid= 18633668)	Maximum peak amp.: 21.073 units (6 times) Time: 220–249 s	Maximum peak amp.: 30.199 units (2 times) Time: 1796, 1820 s
(Breast cancer mammogram image) (By Howard.r.prtty—Own work, CC BY 4.0, https:// commons.wikimedia.org/ w/index.php?curid= 64375620)	Maximum peak amp.: 22.583 units (4 times) Time: 438–454 s	Maximum peak amp.: 31.969 units (6 times) Time: 851–994 s
(Breast cancer mammogram image) (By Unknown photographer—This image was released by the National Cancer Institute, an agency part of the National Institutes of Health, with the ID 2510 (image) (next)., Public Domain, https://commons. wikimedia.org/w/index. php?curid=24052516)	Maximum peak amp.: 47.397 units (18 times) Time: 535–899 s	Maximum peak amp.: 67.037 units (10 times) Time: 464–1799 s

(continued)

Table 2 (continued)

Original image	Z-transform of original image	Z-transform of affine transformed image
(Brain tumour MRI) (By © Nevit Dilmen, CC BY-SA 3.0, https:// commons.wikimedia.org/ w/index.php?curid= 18633690)	Maximum peak amp.: 22.583 units (4 times) Time: 11–20 s	Maximum peak amp.: 25.846 units (2 times) Time: 11, 14 s
(Brain tumour MRI) (https://commons. wikimedia.org/w/index. php?curid=35028568)	Maximum peak amp.: 30.984 units (2 times) Time: 785, 810 s	Maximum peak amp.: 34.684 units (1 time) Time: 943 s
(Breast cancer mammogram image) (By Unknown photographer—This image was released by the National Cancer Institute, an agency part of the National Institutes of Health, with the ID 2703 (image) (next)., Public Domain, https://commons. wikimedia.org/w/index. php?curid=24052431)	Maximum peak amp.:24.382 units (5 times) Time: 41–268 s	Maximum peak amp.: 35.086 units (2 times) Time: 264, 268 s

(continued)

Table 2 (continued)

Original image	Z-transform of original image	Z-transform of affine transformed image
(Breast cancer mammogram image) 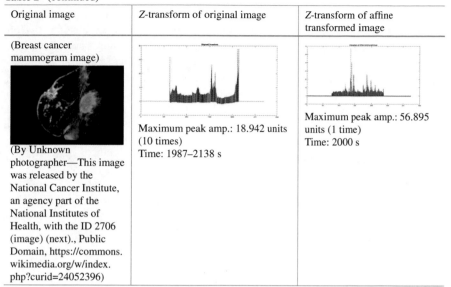 (By Unknown photographer—This image was released by the National Cancer Institute, an agency part of the National Institutes of Health, with the ID 2706 (image) (next)., Public Domain, https://commons. wikimedia.org/w/index. php?curid=24052396)	Maximum peak amp.: 18.942 units (10 times) Time: 1987–2138 s	Maximum peak amp.: 56.895 units (1 time) Time: 2000 s

7 Image Assessment Parameters

There are various features by which we can assess the quality of image which comes as an output. Using these parameters, the overall methodology is compared with the existing ones. Most significant features are true positive (t_p), true negative (t_n), false positive (f_p) and false negative (f_n). True positive means where the conditions are positive and the pixels found correctly. True negative is defined as correct negative prediction where the pixels are found incorrectly, and basically, they do not exist actually. False positive is one type of error and defines as incorrect positive prediction. These pixels do actually not exist, but due to some error of methodology, they come as if a component of output image. False negative is also one type of error which is defined as incorrect negative prediction. The sum of 't_p', 't_n', 'f_p' and 'f_n' is equivalent to total number of subjects to study. The accuracy of the image is defined as ration of sum of 't_p' and 't_n' with the sum of 't_p', 't_n', 'f_p' and 'f_n'. The sensitivity of the output image is defined as the ratio of 't_p' with the sum of 't_p' and 'f_n', whereas the specificity of the image is defined as the ratio of 't_n' and sum of 't_n' and 'f_p'. So we always try to enhance the value of 't_p' and diminish the value of 'f_p'.

We now compare the original image and affine transformed image with respect to image assessment parameters.

8 Conclusion

From Table 3, we can accomplish that true positive, true negative, false positive and false negative are more or less remain the same, because both image's characteristics remain unaffected after affine transformation, so it proves that affine transformation does not affect the image characteristics. The confusion value diminished in the case of affine transformed image rather than original image, which denotes the confusion created to sense the cancerous cell in affine transformed image, is much less than in the case of original image, so we can easily spot the cancerous cell location with less confusion. Next to confusion value, there is four rows titled as true positive rate, true negative rate, false positive rate and false negative rate, which actually predicts

Table 3 Comparison of original images and affine transformed images with respect to image assessment parameters

Parameters	Image 2		Image 4		Image 5	
	Original image	Affine transformed image	Original image	Affine transformed image	Original image	Affine transformed image
True positive (TP)	3.00	3.00	0.01	0.01	0.05	0.05
True negative (TN)	2.00	2.00	1.00	1.00	4.00	4.00
False positive (FP)	0.01	0.01	0.02	0.02	0.01	0.01
False negative (FN)	0.03	0.04	13.00	13.00	0.02	0.02
Confusion value	283.459890	196.192308	0.998866	0.900000	1.996094	0.924531
True positive rate (TPR)	1.666667	2.425965	0.200000	0.333333	0.800000	1.042857
True negative rate (TNR)	727.405530	728.325515	692.999100	692.999093	767.002515	767.002498
False positive rate (FPR)	232.333333	187.574035	238.800000	236.666667	305.200000	300.957143
False negative rate (FNR)	3.594470	2.674485	1.000900	1.000907	0.997485	0.997502
Accuracy	0.78500	1.00500	0.07143	1.37143	1.00000	1.50000
Sensitivity	1.66667	2.42596	0.20000	0.33333	0.80000	1.04286
Specificity	727.40553	728.32552	692.99910	692.99909	767.00251	767.00250

that the image is truthfully cancerous or it is just a false alarm. If we consider the above table, then we can clinch that true positive rate is increased in every case, which denotes that the cancerous cells or some affected cells may be present in the image, which is detected using affine transformation; it leads to that the image is truly positive with the predicted decease. True negative rate is not changed quite because both image's characteristics are same. False positive rate decreases a certain amount for the affine transformed image than original image, which leads to decrease in confusion in the presence of cancerous cell to that specific image when the geometric dimension of the image is altered with the help of affine transformed image. Change in geometric dimension leads to geometrical change of location of affected or cancerous cell. In the case of false negative rate, it is decreased an amount in the case of affine transformed image than original image, which helps to spot the location of cancerous cell. It is noticeable that accuracy and sensitivity are increased in the case of affine transformed image than the original image. Increase in accuracy aids to segment and sense the diseased cells more accurately and precisely without affecting other regular cells. Upsurge in sensitivity is an indication that cancerous cell is changing its physiognomies from benign to malignant; more sensitivity suggests that the cells essential to operate to shield other cells.

References

1. Kumar, R., Srivastava, R., & Srivastava, S. (2015). Detection and classification of cancer from microscopic biopsy images using clinically significant and biologically interpretable features. *Journal of Medical Engineering, 2015*, 457906.
2. Williams, I., Bowring, N., & Svoboda, D. (2014). A performance evaluation of statistical tests for edge detection in textured images. *Computer Vision and Image Understanding, 122*, 115–130.
3. Nadernejad, E., Sharifzadeh, S., & Hassanpour, H. (2008). Edge detection techniques: evaluations and comparisons. *Applied Mathematical Sciences, 2*(31), 1507–1520.
4. Bhardwaja, S., & Mittal, A. (2012). A survey on various edge detector techniques. *Procedia Technology, 4*, 220–226.
5. Bueno, G., González, R., Déniz, O., García-Rojo, M., González-García, J., Fernández-Carrobles, M. M., et al. (2012). A parallel solution for high resolution histological image analysis. *Computer Methods and Programs in Biomedicine, 108*(1), 388–401.
6. Ortiz, A., & Oliver, G. (2006). On the use of the overlapping area matrix for image segmentation evaluation: A survey and new performance measures. *Pattern Recognition Letters, 27*(16), 1916–1926.
7. Zhang, D., & Lu, G. (2004). Review of shape representation and description techniques. *Pattern Recognition, 37*(1), 1–19.
8. Maragos, P. (1989). Pattern spectrum and multiscale shape representation. *IEEE Transactions on Pattern Analysis and Machine Intelligence, 11*(7), 701–716.
9. Flusser, J., & Suk, T. (1993). Pattern recognition by affine moment invariants. *Pattern Recognition, 26*(1), 163–174.
10. Pang, E., & Hatzinakos, D. (1997). An efficient implementation of affine transformation using one-dimensional FFTs. In *IEEE International Conference on Acoustics, Speech, and Signal Processing.*
11. Xie, M. (1995). Feature matching and affine transformation for 2D cell animation. *The Visual Computer, 11*(8), 419–428.

12. Bebis, G., Georgiopoulos, M., Lobo, N. D. V., & Shah, M. (1999). Learning affine transformations. *Pattern Recognition, 32*(10), 1783–1799.
13. Zhao, Y., & Yuan, B. (1998). A new affine transformation: its theory and application to image coding. *IEEE Transactions on Circuits and Systems for Video Technology, 8*(3), 269–274.
14. Lee, S., Lee, G. G., Jang, E. S., & Kim, W. Y. (2006). Fast affine transform for real-time machine vision applications. *Lecture Notes in Computer Science: Intelligent Computing, 4113,* 1180–1190.
15. Zhang, Y. (2009) Image processing using spatial transform. In *IEEE International Conference on Image Analysis and Signal Processing*.
16. Biswal, P. K., & Banerjee, S. (2010). A parallel approach for affine transform of 3D biomedical images. In *IEEE International Conference on Electronics and Information Engineering*.
17. Yang, J., Chen, Y., & Scalia, M. (2012). Construction of affine invariant functions in spatial domain. *Mathematical Problems in Engineering, 2012,* 690262.
18. Biswal, P. K., Mondal, P., & Banerjee, S. (2013). Parallel architecture for accelerating affine transform in high-speed imaging systems. *Journal of Real-Time Image Processing, 8*(1), 69–79.
19. Dai, X., Zhang, H., Liu, T., Shu, H., & Luo, L. (2014). Legendre moment invariants to blur and affine transformation and their use in image recognition. *Pattern Analysis and Applications, 17*(2), 311–326.
20. Hua, D., Li, W. T., & Shi, X. W. (2014). Pattern synthesis for large planar arrays using a modified alternating projection method in an affine coordinate system. *Progress in Electromagnetics Research M, 39,* 53–63.
21. Lee, Y. S., Hwang, W. L., & Tian, X. (2015). Continuous piecewise affine transformation for image registration. *International Journal of Wavelets, Multiresolution and Information Processing, 13*(1), 1550006.
22. Yum, J., Lee, C. H., Park, J., Kim, J. S., & Lee, H. J. (2018). A hardware architecture for the affine-invariant extension of SIFT. *IEEE Transactions on Circuits and Systems for Video Technology, 28*(11), 3251–3261.
23. Liu, N., Pan, J. S., & Nguyen, T. T. (2019). A bi-population quasi-affine transformation evolution algorithm for global optimization and its application to dynamic deployment in wireless sensor networks. *EURASIP Journal on Wireless Communications and Networking, 2019,* 175.
24. Liu, N., Pan, J. S., Wang, J., & Nguyen, T. T. (2019). An adaptation multi-group quasi-affine transformation evolutionary algorithm for global optimization and its application in node localization in wireless sensor networks. *Sensors, 19,* 4112.
25. Fisher, R., Perkins, S., Walker, A., & Wolfart, E. (1996). *Hypermedia image processing reference*. Wiley.

A Comparative Study of Different Feature Descriptors for Video-Based Human Action Recognition

Swarnava Sadhukhan, Siddhartha Mallick, Pawan Kumar Singh,
Ram Sarkar, and Debotosh Bhattacharjee

Abstract Human action recognition (HAR) has been a well-studied research topic in the field of computer vision since the past two decades. The objective of HAR is to detect and recognize actions performed by one or more persons based on a series of observations. In this paper, a comparative study of different feature descriptors applied for HAR on video datasets is presented. In particular, we estimate four standard feature descriptors, namely histogram of oriented gradients (HOG), gray-level co-occurrence matrix (GLCM), speeded-up robust features (SURF), and Graphics and Intelligence-based Scripting Technology (GIST) descriptors from RGB videos, after performing background subtraction and creating a minimum bounding box surrounding the human object. To further speed up the overall process, we apply an efficient sparse filtering method, which reduces the number of features by eliminating the redundant features and assigning weights to the features left after elimination. Finally, the performance of the said feature descriptors on three standard benchmark video datasets *namely,* KTH, HMDB51, and UCF11 has been analyzed.

Keywords Human action recognition · HOG · SURF · GLCM · GIST · KTH · HMDB51 · UCF11

S. Sadhukhan · S. Mallick
Department of Information Technology, Indian Institute of Engineering Science and Technology, Howrah-711103, West Bengal, India
e-mail: sadhu.man11@gmail.com

S. Mallick
e-mail: msiddhartha1606@gmail.com

P. K. Singh (✉)
Department of Information Technology, Jadavpur University, Kolkata-700106, West Bengal, India
e-mail: pawansingh.ju@gmail.com

R. Sarkar · D. Bhattacharjee
Department of Computer Science and Engineering, Jadavpur University, Kolkata-700032, West Bengal, India
e-mail: raamsarkar@gmail.com

D. Bhattacharjee
e-mail: debotoshb@hotmail.com

© Springer Nature Singapore Pte Ltd. 2020
J. K. Mandal and S. Banerjee (eds.), *Intelligent Computing: Image Processing Based Applications*, Advances in Intelligent Systems and Computing 1157,
https://doi.org/10.1007/978-981-15-4288-6_3

1   Introduction

Automatic action recognition of human beings from video, still image, or sensor data has become a benchmark research topic in the field of computer vision. It has various applications in human–computer interactions, surveillance systems, and robotics. Gradually, it has extended its application in fields like health care, video indexing, multimedia retrieval, social networking, and education as well [1–3]. For action recognition, various approaches have been attempted by the researchers over the years. In this paper, our focus is on a specific field of action recognition, i.e., human action recognition (HAR). HAR is the problem of predicting what a person is doing based on a trace of their movements. Movements are often normal indoor activities such as standing, sitting, jumping, and going upstairs. It has become a crucial aspect of human–computer interaction. Its varied range of applications is what drives many researchers across the globe to work in this field. It has important applications that involve domains like annotation and retrieval of video content, health care, and scene modeling to name a few.

Classical machine learning approaches to solve classification problems involve handcrafting features from the input data and building training models using some standard classifiers such as neural network (NN), support vector machine (SVM), K-nearest neighbor (KNN), and random forest. The difficulty is that this feature engineering requires sound expertise in the field. Recently, deep learning methods such as recurrent neural network (RNN) and convolutional neural network (CNN) have been shown to provide state-of-the-art results on challenging activity recognition tasks with little or no feature engineering, instead of using feature learning from raw data [4, 5]. Deep learning methods have been achieving success on HAR problems given their ability to automatically learn higher-order features [6]. However, one most limitation of deep learning models is that they require huge amount of computing resources and data which may not be always available. In such cases, machine learning-based methods remain a viable alternative. Keeping this fact in mind, our study focuses on feature engineering-based HAR techniques following conventional machine learning methodology.

2   Previous Works

In this section, we will discuss some of the previous works done in the domain of video-based HAR. Sharif et al. [7] used histogram of oriented gradients (HOG), local binary pattern (LBP), segmentation-based fractal texture analysis (SFTA) and achieved 99.9% accuracy on the KTH dataset. Initially, preprocessing was performed to remove the existing noise. Then a novel weighted segmentation method was introduced for human extraction prior to feature extraction.

Ullah et al. [8] used a combination of gradient boundary histogram (GBH) and motion boundary histogram (MBH) to attain 62.2% on the HMDB51 dataset. [9]

presented a method to construct dynamic hierarchical trees (DHT), and extracted features using improved dense trajectory (iDT) to yield 70.8% on the HMDB51 dataset.

Singh [10], suggested that some problems can be overcome by placing multiple cameras at multiple viewpoints. The components of their proposed framework are three consecutive modules: (i) detection and positioning of the person's background subtraction, (ii) function extraction, and (iii) final activity were referenced by using a set of hidden Markov models (HMMs). Their technique produced 99.43% accuracy on KTH dataset.

Decent results on the KTH dataset were also obtained by [11], using texture-type descriptors. Chakraborty et al. [12] and Li et al. [13] also used these descriptors for obtaining improved results on the KTH, YouTube, CVC, Hollywood, and Weizmann datasets. [14], Yingying Miao, 2014, used HOG and GLCM feature descriptors and achieved good results. Mahadevan et al. [15] used texture-type feature descriptors such as GLCM for feature extraction.

Yu et al. [16] proposed a method that used semantic texton forests (STF), pyramidal spatiotemporal relationship match (PSRM) on KTH, and UT-Interaction datasets. Using the structure of STFs, the proposed PSRM kernel obtained robust structural matching and avoided quantization effects. A novel fast interest point detector also accelerated their recognition accuracy.

Shabani et al. [17] improved upon salient feature detection by investigating the effect of symmetry and causality of the video filtering. Based on the positive precision and reproducibility tests, they proposed the use of temporally asymmetric filtering for robust motion feature detection and action recognition. They achieved excellent results on Weizmann, KTH, and UCF Sports datasets. Their method provided more precise and more robust features than the Gabor filter. Their temporal filter for asymmetric sync gave better results in case of camera zoom than Gabor filter.

Zhen and Shao [18] used HOG3D with various encoding methods to achieve 94.1% accuracy on the KTH dataset. Sharif et al. [19] presented a framework involving the segmentation and combination of Euclidean distance and of joint entropy-based feature selection. They extracted HOG and Haralick features and used SVMs to get 95.80% accuracy on the Weizmann dataset and 99.30% on the KTH dataset.

Wu et al. [20] proposed a method for HAR that represented each action class by hidden temporal methods. The video segment was described by a temporal pyramid model to capture the temporal structures. For the large intra-class variations, multiple models were combined using OR operation to represent alternative structures. They achieved an overall accuracy of 84.3% on Olympic Sports dataset and 47.1% on HMDB51 dataset. These datasets are very challenging because they employ complex motions, varied camera angles, and also contain camera motion.

In the present work, we have used simple yet comprehensive shape-based and texture-based features for solving the problem of HAR. Our aim is to provide an effective HAR model which takes less computational time, consuming low resource, and also yields good recognition results on video datasets having different number of action classes. It is to be noted that our model yields effective results because of the combination of various feature descriptors used with an effective feature selection method known as sparse filtering.

3   Methods and Methodologies

Our present work on developing a HAR model is mainly divided into six modules— Data preprocessing, Background subtraction, Creation of minimum bounding box, Feature extraction, Feature selection, and finally, Actions classification. Figure 1 shows the workflow of our methodology. Each of the modules is discussed below in detail.

3.1   Data Preprocessing

In order to represent a video in the feature space efficiently, truncation operation on the dataset must be carried out in order to decrease the number of features evaluated from each video and also reduce the computational time. In a typical human activity video, there exist many frames which give unnecessary information such as a frame where the actor is not present or even present partially in the scene. Including such frames in estimating features will reduce the usefulness of the feature descriptor; thereby resulting poor performance of the classification model. In order to ignore this problem, elimination of such unnecessary frames is required in order to increase the overall accuracy of prediction. Two such non-informative video frames are shown in Fig. 2.

Another problem is that adjacent frames do not have significant motion changes of the subject in between them, as shown in Fig. 3. After a few frames, there might be a significant change in motion or spatial structure of the human. Including the frames which are distinguishable from their former frames should be done in order to reduce computational time and reduce the number of features evaluated from each video, as shown in Fig. 4. In order to achieve this, we considered a total of 20 frames from each video datasets at a regular interval. The interval of selecting the frames can be calculated by:

Fig. 1 Schematic diagram of the HAR model understudy

Fig. 2 Frames taken from a video of the class *'running'* from KTH dataset where the subject is: **a** not present in the scene and **b** partially present in the scene

Fig. 3 Images showing two successive frames which do not have any significant change in motion/posture Taken from handclapping action in KTH dataset

Fig. 4 Images showing two frames which have a significant change in motion/posture Taken from *'handclapping'* action in KTH dataset

$$\text{Interval} = \text{floor}\left(\frac{\text{Total number of frames}}{19}\right) \quad (1)$$

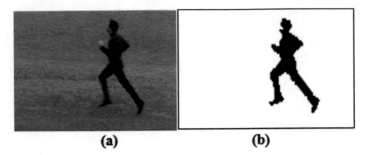

(a)   **(b)**

Fig. 5 Images showing: **a** frame extracted from '*running*' class of KTH dataset and **b** image after applying background subtraction method to the extracted frame shown in (**a**)

3.2 Background Subtraction and Silhouette Extraction

One of the main challenges faced in HAR domain is to remove background occlusion from individual videos. Every shade of color adds dimension to the image, but they are not very much impactful in shape-based HAR model. Thus, removing the background from a frame is required as well as representing the frames by only two levels (binary image) is also desired. The popular method of removing background occlusion is background subtraction and silhouette extraction [21].

There are many standard algorithms which are pretty efficient in extracting human silhouette from a given frame. In our experiment, we have used Otsu's threshold selection method [22] in order to extract silhouette from the selected frame. One example of applying this background subtraction method in one of the classes of KTH dataset is shown in Fig. 5.

3.3 Minimum Bounding Box

Most part of the background subtracted frame does not contain any information contributable for feature extraction as can be seen in Fig. 5(b).

Thus, after extracting silhouette from a given frame of a video it is necessary to detect the minimum bounding box which contains the whole human silhouette. In order to achieve this, the following algorithm can be implemented:

Let the background subtracted image be represented by an image matrix $BSI(i, j)$ where (i, j) is the pixel position of the image matrix. Also let $\text{left}_{BBOX(x)}$ be a 1D array containing all the leftmost coordinated positions of the silhouette. In our case, the silhouette is black and the background is white.

1. For the leftmost part, scan the binary image from left to right using a nested loop iterating i and j:

$$\text{if } BSI(i, j) == 0 \text{ and } BSI(i, j - 1) == 0$$

Fig. 6 a Frame after applying background subtraction and **b** image after applying minimum bounding box to image (**a**)

$$\text{then left}_{\text{BBOX}(x)} = j;$$

$$x + +;$$

break the second nested loop iterating j;

2. Take the minimum of $\text{left}_{\text{BBOX}(x)}$ and save it in p^1;
3. Repeat the same if statement for rightmost part, topmost part, and bottommost part. Scan the image from right to left for obtaining the rightmost part, take the maximum of $\text{right}_{\text{BBOX}(x)}$, and store it in p^2. Scan the image from top to bottom for obtaining the topmost part, take the minimum of $\text{top}_{\text{BBOX}(x)}$, and store it in p^3. Scan the image from bottom to top for obtaining the bottommost part, take the maximum of $\text{bottom}_{\text{BBOX}(x)}$, and store it in p^4.
4. Crop the BSI(i, j) from p^3 to p^4 for obtaining the vertical axis of the minimum bounding box and from p^1 to p^2 for obtaining the horizontal axis of the minimum bounding box.

After applying minimum bounding box for each frame, the human figure performing the action has been resized to a pre-defined dimension of 120 × 60 pixels. This dimension is kept common for all the frames, regardless of the video or dataset it belonged to. An example of applying this method of minimum bounding box to an image of a class of KTH dataset is shown in Fig. 6.

3.4 Feature Extraction

A set of four different texture-based feature descriptors, namely histogram of oriented gradients (HOG), speeded-up robust features (SURF), gray-level co-occurrence matrix (GLCM), and GIST features, have been extracted from the preprocessed videos. These features are calculated first on each of the selected 20 frames after background subtraction and detection of minimum bounding box surrounding the human figure. Then a specific number of feature values from each frame are selected by sparse filtering method (discussed later in the paper). After the selection of the above feature values, these feature values of all the 20 frames are then re-arranged in a row vector one after another which finally represents the whole video. The features extracted from the datasets used for our experiment are mentioned below.

3.4.1 HOG

The HOG is a feature descriptor used in computer vision and also image processing methods for object detection [23]. The HOG feature descriptor counts occurrences of gradient orientation in localized portions of an image. At first, horizontal and vertical gradients are calculated from each minimum bounding box subtracted from the given frames of an input video. This can be easily done by filtering the image with the kernels [-1 0 1] and [-1 0 1]^T. The magnitude and direction of gradient are then calculated for every pixel. The image is divided into small blocks or regions called cells of 16×16 pixels, and then a histogram containing 9 bins corresponding to the angles 0–180 degrees is created. Finally, a set of 8640 features are extracted from each video frame using HOG descriptor.

After calculating HOG features from each of the 20 selected frames, 100 of the most effective features are selected from each frame using sparse filtering technique [24]. The selected features for these 20 frames are then aligned into a row vector, and thus for HOG, a total of 2000 features are selected for each action class.

3.4.2 SURF

SURF is a local feature detector, and it was first proposed by Bay, Tuytelaars, and Gool L Van in their work (Bay et al. [25]. It is known to be an upgraded version of scale-invariant feature transform (SIFT). In SIFT, Laplacian of Gaussian is approximated with box filter by SURF feature descriptor. The method is very fast because the detection technique uses an integral image where the value of a pixel at (x, y) is the summation of all values in the rectangle. For detection of features the Hessian matrix $H = \begin{bmatrix} L_{xx} & L_{xy} \\ L_{yx} & L_{yy} \end{bmatrix}$, where L_{xx} is the convolution of the second derivative of a Gaussian with the image at the point (x, y). Valid features are found as local maxima of the Hessian determinant values over range of $3 \times 3 \times 3$ where the third dimension represents the window size.

The SURF descriptor is scale and rotation invariant. To make the descriptor scale invariant, it is sampled over a window which is proportional in scale with the detected window size. In order to achieve rotational invariance, the orientation of the point of interest needs to be found. The Haar wavelet responses in both x- and y-directions within a circular neighborhood of radius 6S around the point of interest are computed, where S is the scale at which the point of interest was detected. The obtained responses are weighted by a Gaussian function centered at the point of interest, then plotted as points in a two-dimensional space, with the horizontal response in the abscissa and the vertical response in the ordinate. After rotation, to describe the region around the point, a square region is extracted, centered on the interest point, and oriented along the orientation as selected above. The size of this window is 20S.

The interest region is split into smaller 4×4 square subregions, and for each one, the Haar wavelet responses are extracted at 5×5 regularly spaced sample points. The responses are weighted with a Gaussian filter. Finally, a feature vector of dimension 400 is designed using the SURF feature descriptor.

After calculating SURF features from each of the 20 selected frames, 20 feature values are selected from each frame using sparse filtering technique [24]. The selected features for these 20 frames are then aligned into a row vector, and thus for SURF feature, a total of 400 feature values are selected for each action class.

3.4.3 GLCM

GLCM was first proposed by Haralick et al. [26] in his work 'Textural Features for Image Classification' [26]. GLCM with associated texture features is a method of extracting statistical texture features of second order.

GLCM is a matrix where the number of rows and columns is equal to the gray levels in the image. The matrix element P or P_{ij} is the conditional probability of the intensities of two pixels with intensities i and j given that the separation between the two pixels is $(\text{del}_x, \text{del}_y)$. On the other hand, the matrix element $P(i, j|d, \theta)$ is the second-order statistical probability for change between i and j for a given angle θ and distance d.

From the second-order statistics mentioned above, one can derive some useful properties, and all of them have been used for our experiment. There are a total of 13 features calculated from GLCM, each feature containing a definite amount of information about an input image. These features are as follows: energy, entropy, homogeneity, correlation, etc. GLCM also requires an 'offset' value which defines relationships of varying direction, and distance and 'offset' values of $[01; -11; -10; -1 - 1]$ (represented in a matrix form) are used for calculating the feature values from each selected frame. As a result, a feature set consisting of 1040 elements is produced.

After calculating the GLCM-based features from each of the 20 selected frames, 50 feature values are selected from each frame using sparse filtering technique [24]. The selected feature values for these 20 frames are then aligned into a row vector, and thus for GLCM feature, a total of 1000 feature values are selected for each action class.

3.4.4 GIST Features

GIST features are used for estimating various statistical characteristics of an image [27]. GIST features are calculated by using convolution of the Gabor filter with each of the selected frames at different scales and orientation. In our experiment, we have used a Gabor filter transfer function having orientation per scale as [8 8 8]. The values for each convolution of the filter at each scale and orientation are used as GIST features for a particular frame. Thus, in this way, repetitive gradient directions having both low and high frequencies of a frame can be measured. GIST feature extraction filters the input frame first into a number of low-level visual feature channels such as color, orientation, motion, intensity, and flicker. Within each sub-channel, center-surround operations are performed between filter output maps at different scales. With the above feature map, a saliency map can be yielded by using the feature map to detect clearly visible regions in each channel. Now each sub-channel extracts a GIST vector from its corresponding feature map after computing the low-level features. A 4 × 4 grid is used in our experiment for averaging the conspicuous subregions over the map which outputs a feature vector of size 6400 for each video frame.

After calculating GIST features from each of the 20 selected frames, 100 features are selected from each frame using sparse filtering technique [24]. The selected features for these 20 frames are then aligned into a row vector, and thus for gist feature, a total of 2000 features are selected for each action class.

4 Feature Selection

Feature selection is a method that map input features to select a set of new output feature subsets by eliminating irrelevant and redundant features. Many feature extraction methods use unsupervised learning to extract features. Unlike some feature extraction methods such as principal component analysis (PCA) and non-negative matrix factorization (NNMF), sparse filtering [24] can increase dimensionality (or decrease dimensionality) of the total feature values derived from the feature descriptors of each frame.

In sparse filtering, the sparsity in the feature distribution is optimized exclusively. This method avoids explicit modeling of the data distribution giving rise to a simple formulation and permits efficient learning. Moreover, it is a hyperparameter-free approach which implies that sparse filtering works well on a range of data modalities (whether the data is structured, semi-structured, or unstructured) without the need for specific tuning of modal parameters.

In our work, we have reduced the total number of features per frame through sparse filtering into a feasible total of 5400 features (2000 features for HOG, 400 features for SURF, 1000 features for GLCM, and 2000 features for GIST) representing a video as mentioned in each section of the feature descriptors. At first, sparse filtering is applied on all the frames while estimating each feature descriptor, and then the total number of features is reduced into a specific number of features for each feature descriptor (2000 features for HOG, 400 features for SURF, 1000 features for GLCM, and 2000 features for GIST). All the features of each feature descriptor are arranged in four row vectors (one row vector for HOG feature descriptor, one row vector for SURF, and so on), and each row vector contains the features of one feature descriptor applied on all the 20 frames together. Now each feature descriptor row matrix, each containing the feature values of all the frames is appended one after the other. Thus, all the four features are fused one after the other and the resulting row matrix represents the feature values for a particular action video class.

5 Experiments and Analysis

We have worked on three HAR datasets, namely KTH dataset [28], HMDB 51 [29], and UCF 11 (YouTube dataset) [30]. A brief description about these datasets is presented below:

The KTH dataset [28] is a frequently used dataset which covers 600 video clips. It has six classes of human actions: '*Walking*', '*Running*', '*Jogging*', '*Boxing*', '*Hand waving*', and '*Handclapping*'. Each action class has 100 sequences with 25 actors performing the actions in four different situations: outdoors, outdoors with different outfits, indoors with variations of lighting, and outdoors with variations in scale. The video resolution is 120×160 and the frame rate is 25.

The UCF11 sports action dataset [30] has 1500+ sequences and 11 classes of sports actions like '*Diving*', '*Kicking*', '*Weight-lifting*', '*Horse riding*', '*Golf swinging*', '*Running*', '*Skating*', '*Shooting a baseball ball*', and '*Walking*'. The resolution of videos is 480×720 pixels.

The HMDB51 dataset [29] contains 51 distinct types with at least 101 video clips in each type, which makes a total of 6849 action samples extracted from various range of sources. The categories of actions can be grouped into five types:

- **General facial actions**: smile, chew, laugh, talk
- **Facial action with object interaction**: smoke, eat, drink
- **General body movements**: walk, wave, clap hands, jump, cartwheel, climb, climb stairs, dive, handstand, pull up, push up, sit, fall on the floor, sit up, somersault, turn, stand up, backhand flip.
- **Body movements with object interaction**: brush air, catch, draw sword, golf, hit something, dribble, kick a ball, pick, pour, push something, ride a bike, ride a horse, shoot ball, shoot bow, shoot gun, throw, swing baseball bat, sword exercise.
- **Body movements with human interaction**: fencing, hugging, kicking somebody, kissing, punching, shaking hands, sword fighting.

We have evaluated our HAR model using three standard classifiers, namely SVM [31], KNN [32], and RF [33]. A threefold cross-validation is used for the evaluation. The evaluation metric used for measuring the performance of the model is classification accuracy, defined as follows:

Table 1 Classification accuracies achieved by three different classifiers on KTH [28], UCF11 [30], and HMDB51 [29] datasets

Dataset	Number of training action samples	Number of testing action samples	Classification accuracy (%) using		
			SVM (%)	KNN (%)	RF (%)
KTH	400	200	**99.49**	99.16	98.83
UCF11	774	386	**91.59**	85.62	89.81
HMDB51	4666	2334	**70.3**	64.77	65.32

The bold values signifies the number of action videos succeessfully classified for a particular action class

$$\text{Classification Accuracy}(\%) = \frac{\#\text{videos successfully classified}}{\#\text{total videos present}} \times 100 \qquad (2)$$

The classification accuracies obtained using three different classifiers on the above-mentioned datasets are shown in Table 1. It can be seen from Table 1 that SVM classifier attains the highest accuracies of 99.49%, 73.39%, and 70.3% on KTH, UCF11, and HMDB51 datasets, respectively. The confusion matrices obtained for this best case of SVM classifier for KTH and UCF11 datasets are shown in Tables 2 and 3, respectively.

From Table 2, it can be realized that except '*Handclapping*' class, all other classes give 100% accuracy on test set.

The '*Jogging*', '*Running*', and '*Walking*' action classes involve the displacement of the human. This helps in recognizing the actions. The difference in displacements among them for the same interval of frames helps classifying them, whereas a few videos of '*Handclapping*' are not classified correctly due to similar hand movements with '*Boxing*' and '*Waving*'. They also have no displacement. So, factors to classify have got reduced.

From Table 3, it can be inferred that the action classes '*Diving*' and '*Golf swinging*' show the best results. '*Golf swinging*' has a very distinct action movement and does

Table 2 Confusion matrix obtained using SVM classifier for KTH dataset

Predicted class	Actual class					
	Boxing	*Handclapping*	*Hand waving*	*Jogging*	*Running*	*Walking*
Boxing	**100**	0	0	0	0	0
Handclapping	3	**96**	0	1	0	0
Hand waving	0	0	**100**	0	0	0
Jogging	0	0	0	**100**	0	0
Running	0	0	0	0	**100**	0
Walking	0	0	0	0	0	**100**

The bold values signifies the number of action videos succeessfully classified for a particular action class

Table 3 Confusion matrix obtained using SVM classifier for UCF11 dataset

Predicted class	Actual class										
	Soccer juggling	Swinging	Tennis swinging	Trampoline jumping	Walking a dog	Cycling	Diving	Golf swinging	Basketball shooting	Horse Riding	Volleyball spiking
Soccer juggling	**144**	2	1	2	2	0	0	0	1	4	0
Swinging	0	**128**	0	2	1	2	0	0	1	3	0
Tennis swinging	1	2	**152**	1	0	0	3	2	1	3	2
Trampoline jumping	0	1	0	**109**	0	2	0	0	1	5	1
Walking a dog	0	1	4	0	**107**	1	1	1	0	7	1
Cycling	1	1	2	2	2	**124**	0	0	0	13	0
Diving	2	0	0	0	0	2	**133**	1	0	1	0
Golf swinging	0	1	1	1	0	0	2	**136**	1	0	0
Basketball shooting	0	2	1	1	0	1	3	1	**125**	2	0
Horse riding	0	2	2	1	0	5	1	1	2	**175**	2
Volleyball spiking	1	1	0	0	1	1	1	0	2	3	**106**

The bold values signifies the number of action videos successfully classified for a particular action class

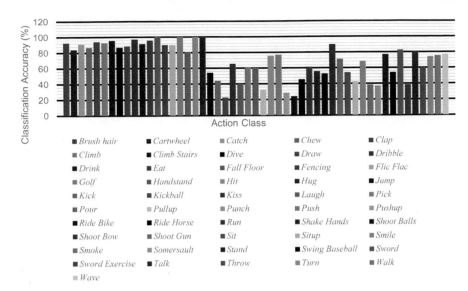

Fig. 7   Class-wise performance achieved by SVM classifier on HMDB51 dataset

not vary much among different videos. This accounts for its good accuracy. However, the '*Diving*' class includes a very distinct movement within the periphery of the video, i.e., dropping from a height and diving into a swimming pool. Many videos are misclassified as '*Horse Riding*' as the action class '*Horse Riding*' has a total of 191 videos which is the highest among all the action classes in the confusion matrix. Thus, this might have led to overfitting and reduced the overall accuracy of prediction.

From Fig. 7, it is evident that the action classes like '*Falling on the floor*', '*golf*', '*hit*', '*hug*', and '*dribble*' are classified correctly. Also, some action classes like '*hug*' and '*dribble*' have more than one action figure in the videos. This made it easier for the classifiers to classify them. Some action classes like '*golf*', '*hit*', and '*falling on the floor*' have a distinct action and do not vary much with change in videos. Thus, they are predicted with certainty. On the other hand, the classes like '*Kickball*', '*Somersault*', and '*Riding bike*' have yielded low accuracies. This has been a result of misclassification as they were confused with some other classes with similar actions or because some other similar classes got overtrained by the classifier. This is a common problem that happens with large datasets having some similar action classes.

In the present work, we have compared our results with some existing HAR models developed using machine learning-based methods. Table 4 shows the comparison results on the three datasets.

Table 4 Comparison of the proposed work with some previous HAR models evaluated on KTH, UCF11, and HMDB51 datasets

Dataset	Authors	Method	Classification accuracy (%)
KTH [28]	Kim et al. [34]	Motion segment decomposition and model matching	95.00
	Chakraborty et al. [12]	BoVW model of N-jet features	96.35
	Sahoo [11]	Bag of histogram of optical flow	96.70
	Zhou and Zhang [35]	HMM	98.33
	Singh [10]	HMM and LBP	99.43
	Proposed Model	**HOG+SURF+GLCM+GIST**	**99.49**
UCF11 [30]	Chen and Grauman [36]	Max sub-graph search	55.04
	Yuan et al. [37]	Bag of visual words with R transform	87.33
	Yan and Luo [38]	AdaBoost and sparse representation	90.67
	Harshalla (2017)	CNN and LSTM	94.60
	Proposed Model	**HOG+SURF+GLCM+GIST**	**91.59**
HMDB51 (Kuehne and Jhuang 2011)	Wu et al. [20]	Hidden temporal methods	47.1
	Ullah et al. [8]	GBH and MBH	70.3
	Wang et al. [9]	Dynamic hierarchical trees	70.8
	Proposed Model	**HOG+SURF+GLCM+GIST**	**70.3**

The bold values signifies the number of action videos successfully classified for a particular action class

6 Conclusion

In this paper, we have implemented a comparative study of four different feature descriptors for performing action recognition from three standard video datasets, namely KTH, UCF11, and HMDB51. Moreover, a sparse filtering-based technique is also used for reducing the size of original feature vector. Our study, when tested on three benchmark datasets, namely KTH, HMDB51, and UCF11, produced recognition accuracies of 99.49%, 70.3%, and 91.59%, respectively. It is to be noted that our method has the ability to classify human actions on highly unconstrained videos and also on datasets having large number of action classes. Our approach is entirely based on machine learning-based methods. We have used supervised classifiers like SVM, KNN, and RF for evaluating the classification accuracies for HAR problem.

7 Future Scope

HAR has been used for many essential applications in scientific, commercial, and domestic domains. But there are other areas where HAR can be implemented. Some are outlined below:

- **Surveillance System**—HAR can be used in video surveillance in commercial areas, banks, ATMs, railway/metro stations, etc., to detect unnatural human activity and ensure the security of the place.
- **Social Media Platform**—With the development of multimodal approaches, the performance of social media platforms can be improved where human choices, behavior, and moods can be predicted and dealt with accordingly.
- **Complex and Overlapping Activities**—More research work should be carried out to process real-life videos, videos of longer durations, multiple objects in a video, and where more than one action have to be recognized in a particular scene.
- **Better feature selection methods and classification models**—Application of feature selection techniques in the domain of HAR has not explored much. Therefore, more research works need to be done, where various filter, wrapper, or embedded feature selection techniques can be used. The result of a classifier is also relative, as the accuracy varies with the datasets.

References

1. Jaimes, A., & Sebe, N. (2005). Multimodal human computer interaction: A survey. In *International workshop on human-computer interaction* (pp. 1–15). Berlin, Heidelberg: Springer.
2. Osmani, V., Balasubramaniam, S., & Botvich, D. (2008). Human activity recognition in pervasive health-care : Supporting efficient remote collaboration. *Journal of Network and Computer Applications, 31*, 628–655. https://doi.org/10.1016/j.jnca.2007.11.002
3. Jain, A. K., & Li, S. Z. (2011). *Handbook of face recognition* (Vol. 1). New York: springer.
4. Du, Y., Wang, W., & Wang, L. (2015). Hierarchical Recurrent Neural Network for Skeleton Based Action Recognition. In *Proceedings of the IEEE Computer Society Conference on Computer Vision and Pattern Recognition*, 1110–1118.
5. Ji, S., Xu, W., Yang, M., & Yu, K. (2012). 3D convolutional neural networks for human action recognition. *IEEE Transactions on Pattern Analysis and Machine Intelligence, 35*(1), 221–231.
6. Gammulle, H., Denman, S., Sridharan, S., & Fookes, C. (2017, March). Two stream lstm: A deep fusion framework for human action recognition. In *2017 IEEE Winter Conference on Applications of Computer Vision (WACV)* (pp. 177–186) IEEE. https://doi.org/10.1109/WACV. 2017.27
7. Sharif, M., Attique, M., Farooq, K., Jamal, Z., Shah, H., & Akram, T. (2019). Human action recognition : A framework of statistical weighted segmentation and rank correlation-based selection. *Pattern Analysis and Applications*, (0123456789). https://doi.org/10.1007/s10044-019-00789-0
8. Ullah, A., Muhammad, K., Ul, I., & Wook, S. (2019). Action recognition using optimized deep auto encoder and CNN for surveillance data streams of non-stationary environments. *Future Generation Computer Systems, 96*, 386–397. https://doi.org/10.1016/j.future.2019.01.029.

9. Wang, T., Duan, P., & Ma, B. (2019). Action recognition using dynamic hierarchical trees. *Journal of Visual Communication and Image Representation, 61,* 315–325.
10. Singh, R., Kushwaha, A. K. S., & Srivastava, R. (2019). Multi-view recognition system for human activity based on multiple features for video surveillance system. *Multimedia Tools and Applications, 78*(12), 17165–17196.
11. Sahoo, S. P., Silambarasi, R., & Ari, S. (2019, March). Fusion of histogram based features for Human Action Recognition. In *2019 5th International Conference on Advanced Computing & Communication Systems (ICACCS)* (pp. 1012–1016). IEEE.
12. Chakraborty, B., Holte, M. B., Moeslund, T. B., Gonz, J., & Roca, F. X. (2011). *A Selective Spatio-Temporal Interest Point Detector for Human Action Recognition in Complex Scenes.* (pp. 1776–1783).
13. Li, B., Ayazoglu, M., Mao, T., Camps, O. I., & Sznaier, M. (2011). Activity Recognition using Dynamic Subspace Angles. In *IEEE Computer Society Conference on Computer Vision and Pattern Recognition.* https://doi.org/10.1109/CVPR.2011.5995672
14. Rao, A. S., Gubbi, J., Rajasegarar, S., Marusic, S., & Palaniswami, M. (2014). Detection of anomalous crowd behaviour using hyperspherical clustering. In *2014 International Conference on Digital Image Computing: Techniques and Applications (DICTA).* https://doi.org/10.1109/DICTA.2014.7008100.
15. Mahadevan, V., Li, W., Bhalodia, V., & Vasconcelos, N. (2010). Anomaly detection in crowded scenes. In *IEEE Computer Society Conference on Computer Vision and Pattern Recognition (IEEE).* https://doi.org/10.1109/CVPR.2010.5539872
16. Yu, T., Kim, T., & Cipolla, R. (2010). Real-time action recognition by spatiotemporal semantic and structural forests. *British Machine Vision Conference, BMVC, 52*(1-52), 12. https://doi.org/10.5244/C.24.52.
17. Shabani, A. H., Clausi, D. A., & Zelek, J. S. (2011). Improved spatio-temporal salient feature detection for action recognition. *British Machine Vision Conference, 1,* 1–12.
18. Zhen, X., & Shao, L. (2016). Action recognition via spatio-temporal local features: A comprehensive study. *Image and Vision Computing, 50,* 1–13. https://doi.org/10.1016/J.IMAVIS.2016.02.006.
19. Sharif, M., Khan, M. A., Akram, T., Javed, M. Y., Saba, T., & Rehman, A. (2017). A framework of human detection and action recognition based on uniform segmentation and combination of Euclidean distance and joint entropy-based features selection. *Eurasip Journal on Image and Video Processing, 2017*(1). https://doi.org/10.1186/s13640-017-0236-8
20. Wu, J., Hu, D., & Chen, F. (2014). Action recognition by hidden temporal models. *The Visual Computer, 30*(12), 1395–1404. https://doi.org/10.1007/s00371-013-0899-9
21. Elgammal, A., Duraiswami, R., Harwood, D., & Davis, L. S. (2002). Background and foreground modeling using nonparametric kernel density estimation for visual surveillance. *Proceedings of the IEEE, 90*(7), 1151–1163.
22. Otsu, N. (1979). A Threshold selection method from Gray-Level Histograms. *IEEE transactions on systems, Man Cybernetics, 9*(1), 62–66. Retrieved from http://web-ext.u-aizu.ac.jp/course/bmclass/documents/otsu1979.pdf.
23. Dalal, N., & Triggs, B. (2010). Histograms of oriented gradients for human detection. In *IEEE Computer Society Conference on Computer Vision and Pattern Recognition,* (pp. 886–893).
24. Ngiam, J., Chen, Z., Bhaskar, S. A., Koh, P. W., & Ng, A. Y. (2011). Sparse filtering. In *Advances in neural information processing systems* (pp. 1125–1133). Retrieved from https://papers.nips.cc/paper/4334-sparse-filtering.pdf
25. Bay, H., Tuytelaars, T., & Gool, L. Van. (2008). SURF : Speeded up robust features. *Computer Vision and Image Understanding, 110*(3), 346–359. Retrieved from http://www.cescg.org/CESCG-2013/papers/Jakab-Planar_Object_Recognition_Using_Local_Descriptor_Based_On_Histogram_Of_Intensity_Patches.pdf.
26. Haralick, R. M., Shanmugam, K., & Dinstein, I. H. (1973). Textural features for image classification. *IEEE Transactions on Systems, Man, and Cybernetics, SMC-3*(6), 610–62
27. Oliva, A., & Torralba, A. (2001). Modeling the shape of the scene: A holistic representation of the spatial envelope. *International Journal of Computer Vision, 42*(3), 145–175. https://doi.org/10.1023/A:1011139631724.

28. Christian, S., Barbara, C., & Ivan, L. (2004). Recognizing Human Actions : A Local SVM Approach. In *Proceedings of the 17th International Conference on Pattern Recognition, 2004. ICPR 2004* (pp. 3–7).
29. Kuehne, H., Jhuang, H., Garrote, E., Poggio, T., & Serre, T. (2011). HMDB : A large video database for human motion recognition. In *2011 International Conference on Computer Vision.* https://doi.org/10.1109/ICCV.2011.6126543
30. Liu, J., Luo, J., & Mubarak, S. (2009). Recognizing realistic actions from videos " in the wild." In *2009 IEEE Conference on Computer Vision and Pattern Recognition*, (pp. 1–8). https://doi.org/10.1109/CVPR.2009.5206744
31. Evgeniou, T., & Pontil, M. (2011). Support vector machines : Theory and applications Workshop on support vector machines: Theory and applications. In *Machine Learning and Its Applications: Advanced Lectures.* https://doi.org/10.1007/3-540-44673-7
32. Altman, N. S. (1992). An introduction to kernel and nearest-neighbor nonparametric regression. *The American Statistician, 46*(3), 175–185.
33. Ho, T. K. (1998). The random subspace method for constructing decision forests. *IEEE Transactions on Pattern Analysis and Machine Intelligence, 47,* 832–844.
34. Kim, H. J., Lee, J. S., & Yang, H. S. (2007). Human action recognition using a modified convolutional neural network. In *International Symposium on Neural Networks* (pp. 715–723). Berlin, Heidelberg: Springer.
35. Zhou, W., & Zhang, Z. (2014). Human action recognition with multiple-instance markov model. *IEEE Transactions on Information Forensics and Security, 9,* 1581–1591.
36. Chen, C. Y., & Grauman, K. (2016). Efficient activity detection in untrimmed video with max-subgraph search. *IEEE Transactions on Pattern Analysis and Machine Intelligence, 39*(5), 908–921.
37. Yuan, C., Li, X., Hu, W., Ling, H., & Maybank, S. (2013). 3D R transform on spatio-temporal interest points for action recognition. In *Proceedings of the IEEE Conference on Computer Vision and Pattern Recognition* (pp. 724–730). https://doi.org/10.1109/CVPR.2013.99
38. Yan X & Luo Y (2012). Recognizing human actions using a new descriptor based on spatial-temporal interest points and weighted-output. *Neurocomputing, 87.*

Pose Registration of 3D Face Images

Koushik Dutta, Debotosh Bhattacharjee, and Mita Nasipuri

Abstract In computer vision, face analysis is a hot topic in various domains. It is very challenging to establish a robust face recognition system, including different variations of faces such as expression, pose, and occlusion. The problem arises when the system considers any pose-variant images. Pose variation issues are not handled correctly in 2D domain, whereas 3D data easily handles the pose variation issue. There exist different registration techniques for aligning 3D pose-variant data. In this chapter, we have discussed the need for registration, followed by different registration techniques. Before registration, the pose should be detected in any system, so we have discussed some new pose detection techniques. After that, we have given a brief description of currently proposed face registration techniques; those are mainly worked with 3D face data. Experimentally we have illustrated the importance of registration in any face recognition system. We have considered some of the well-known 3D face databases for validation of the mentioned techniques and also put forward some discussions on the relevant issues.

Keywords Face analysis · Pose-variant image · Registration · Pose detection · 3D face registration

1 Introduction

From the perspective of security concerns in the current scenarios of the real world, face analysis is very much critical in different aspects. Face detection [1] and recognition [2] are the two well-known face analysis-based works. Other than these two analyses, facial expression analysis [3], physiognomy analysis [4], face blindness,

K. Dutta (✉) · D. Bhattacharjee · M. Nasipuri
Computer Science and Engineering, Jadavpur University, Kolkata, India
e-mail: koushik.it.22@gmail.com

D. Bhattacharjee
e-mail: debotoshb@hotmail.com

M. Nasipuri
e-mail: mitanasipuri@gmail.com

© Springer Nature Singapore Pte Ltd. 2020
J. K. Mandal and S. Banerjee (eds.), *Intelligent Computing: Image Processing Based Applications*, Advances in Intelligent Systems and Computing 1157,
https://doi.org/10.1007/978-981-15-4288-6_4

and annotation of faces [5] are also important in different fields of real-time applications. Some of the real-time applications of face recognition are like payments, security, law enforcement, advertising, and health care. All the application-based systems can be meaningful or robust when it follows different variations of faces as inputs like expression, occlusion, pose, and combined variations.

In computer vision, face recognition is an active task. The face recognition system can be useful when it follows different pose variation faces as inputs for testing the system. In the case of the 2D face recognition system, the problem of low recognition rate occurs when illumination and pose variation of face images are used as inputs. These problems have easily handled by 3D faces, where the 3D data consists of extra depth information, and the problem of extreme pose variation of faces, i.e., across 90°, can be solved by registering/aligning the pose-variant face to reference frontal face model. It is an initial problem for the researchers to develop any face analysis-based system. The accuracy of recognition in terms of registration error is very much crucial for establishing system in any real-world environment. The variation of the accuracy of face recognition depends on registration accuracy, which is shown through experiments in an upcoming section of this chapter. The training-based face recognition system can handle misalignment of faces, but it would be problematic for local feature-based applications, though some global feature-based approaches like local binary pattern (LBP), principal component analysis (PCA), and Gabor wavelet features can easily handle small alignment-based faces. Other face analysis-based systems, the registration is also essential in various clinical applications [6]. There exist two main categories of registrations as rigid and non-rigid/deformable registrations. In the medical research domain, deformable registration is mostly used in various applications like radiation therapy [7, 8], physical brain deformation in neurosurgery, and lung motion estimation with deformable registration [9]. The registration is also applicable to heritage reconstruction and industrial application.

After the acquisition stage, other than noise removal, the registration/alignment is one of the essential preprocessing stages. The registration is a process of transforming set of points into a single coordinate. Generally, X-axis registration process needs two objects of any specific domain. First, the source object is denoted by $S \in \mathbb{R}^d$ and target object by $T \in \mathbb{R}^d$, $d = \{2, 3\}$. Now, the generalized expression of registration process can be written as in Eq. 1.

$$\widehat{T} = \frac{\text{argmax}}{F \in \mathbb{F}} M(S, T, F) \tag{1}$$

where F denotes transformation function, $\mathbb{F}$ denotes the search space, M denotes similarity measure, $\widehat{T}$ denotes the solution, and Arg max is the optimization.

The rest of the chapter is divided into some significant sections and subsections. Section 2 illustrates the details of pose orientation of the 3D face. A brief review of pose detection techniques is discussed in Sect. 3. The categories of different registration techniques are briefly described in Sect. 4. Section 5 gives a detailed history of 3D face registration research. Next, Sect. 6 discusses the results. Section 7 concludes the chapter.

2 Details of Pose Orientation of 3D Face

In 3D, three basic orientation of an object is possible, according to X-axis, Y-axis, and Z-axis.

Similarly, a human face can be oriented according to the X-axis denoted as roll, Y-axis denoted as yaw, and Z-axis denoted as pitch. Other than this, some of the mixed orientations like X- and Y-axes together also exist in some of the 3D face databases. There exist various well-known and published 3D databases such as Frav3D, Bosphorus3D, Casia3D, and GavabDB. We can categorize the 3D data in two ways: (1) synthetic data (from 2D image to 3D construction) and (2) real data (scanned data using 3D acquisition device). There exist various methods for creation of synthetic 3D data. On the other way, the depth cameras like structured light scanner, laser scanner, Kinect 3D scanner, which are used for acquiring real 3D data. Table 1 illustrates detailed descriptions of pose-variant images of different databases; those are captured using different 3D scanners. Next, Fig. 1 shows different pose orientations of Frav3D face database. After getting the 3D data, generally, it can be represented in three different ways: range or depth or 2.5D image, 3D point clouds in 3D coordinates, and 3D mesh representation. Figure 2 shows the different representations of 3D face data.

3 Review on 3D Pose Detection Techniques

Before registration of 3D faces, pose detection is one of the significant tasks for real-time applications of face analysis. From the last few decades, various techniques of head pose detection have been established by the researchers. Here, we have reviewed some new pose detection techniques.

In [10], the authors have detected the head pose by using an adaptive 3D matched filter. Further, the combination of particle swarm optimization (PSO) and iterative closest point (ICP) algorithms has been used for registering a morphable face model to the measured depth data. Next, in [11], the random regression forest technique is used to estimate the 3D head pose in real-time from depth data. It learns a mapping between depth features and some parameters like 3D head position and rotation angle. They extended the regression forest to discriminate the depth patches for predicting pose and classification of the head. In [12], the authors have used convolutional neural network (CNN)-based framework named as POSEidon+ that exploits the recent deterministic conditional generative adversarial network (GAN) model that generates grayscale image from depth data for head pose estimation. In [13], the authors have used a geometrical model for finding out the pose orientation. They

Table 1 Description of pose-variant images of different databases

3D database	Number of subjects	No. of pose-variant data	Along X-axis (Roll)	Along Y-axis (Yaw)	Along Z-axis (Pitch)	Mixed
Frav3D	106	8	2	4 (+25°, +30°, −25°, −30°)	2	NA
Bosphorus3D	105	13	NA	7 (+10°, +20°, +30°, +45°, +90°, −45°, −90°)	4 (Strong up, Slight up, Strong down, Slight down)	2 (45° Yaw and + 20° Pitch, 45° Yaw and −20° Pitch)
Casia3D	123	20	4 [left and right (20°–30°)]	12 [+ve and −ve (20°–30°, 50°–60°, 80°–90°)]	4 [Up and down (20°–30°)]	NA
GavabDB	61	4	NA	2 (+90° and −90°)	2 (+35° and −35°)	NA

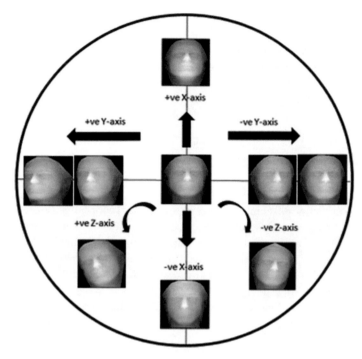

Fig. 1 Variation of face pose

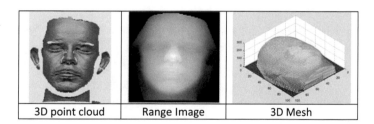

Fig. 2 Different representations of 3D face data

have detected pose from both single and combined axis rotated faces. Initially, nose tip and eye corner landmarks are detected on different pose-variant 3D faces. In the case of single-axis rotation, first, they segregate Z-axis rotated face images from other pose-variant faces by using a horizontal line with respect to eye corners. In case of Z-axis rotated faces, a horizontal line does not exist between two eye corners. Further, for segregating X- and Y-axis, use nose-tip point. In X-axis-based rotated faces, the deviation of nose tip with respect to Y-axis is more than X-axis. Similarly, vice versa for Y-axis-based rotated faces. Now, for combined/composite-oriented faces, they have checked previous methods of any two pose detection in X-, Y-, and Z-axes rotation together. In [14], the authors use bilateral symmetry of face and estimate the pose based on central profile. The central profile is intersection curve between a symmetry plane and 3D surface. The curve starts from forehead and ends up to chin portion. The normal from each point of the curve is parallel to symmetry plane. The work generates a symmetry plane from the central profile curve. The pose detected based on deviation angle of symmetry plane. Currently, most of the researchers have used CNN-based techniques for head pose detection from 2D face data.

4 Registration Technique

Registration is an essential preprocessing operation of any real-time face analysis system that aligns the different 3D views into a standard coordinate system. In the case of a face recognition system, registration is a crucial step for overall performance of recognition. There are various techniques of registration in 2D and 3D domains. In 3D registration, always full model reconstruction has been done. So, considering efficient technique for taking less amount of time, when massive number of point clouds will be registered. After the registration, there are no such optimized techniques for matching. The registration techniques can be divided into three categories as a whole face, point, and part registrations.

4.1 Whole Face Registration

In this technique, the registration is done using the full-face region. Further, it can be divided into two distinct parts as rigid and non-rigid.

Rigid Registration Rigid registration is the global transformation [15] through some landmark points. It is also called Euclidean or affine transformation. It includes rotations, translations, reflections, and their combinations. This registration technique is very much simple compared to others. Generally, most of the system used eye, nose, and mouth position-based landmarks [16, 17] for transformation. In some of the cases, many points (nearly 50–60) have used for transformation; it is also termed as active appearance model (AAM)-based [18] transformation. These model-based techniques are sensitive to registration error compared to the landmarks-based approaches. Some traditional rigid registration techniques [19–21] register the surfaces using a 3D point cloud. An iterative algorithm is a beneficial algorithm for rigid displacement between two separate point clouds of the surfaces.

Iterative Closest Point This is one of the most popular rigid registration techniques in any working domain for 3D data. The iterative algorithm is used for superposing two surfaces, S_1 on S_2, which is the rigid displacement (R, t) of one surface, where R denotes orthogonal transformation, and t is the translating vector of the origin. Each iteration i of the iterative algorithm consists of two steps. The details of the algorithm are given below.

Algorithm ICP

Step 1: For each points P of S_1, a pair (P, Q) is added to $Match_i$, Q is the closest points of S_2 to the point $R_{(i-1)}*P + t_{(i-1)}$. The distance map method [22] is generally used for computing the closest point.
Step 2: In this step, the error metric calculation is done for accurate displacement.
Generally, the least squares method [23] is used for error calculation.

The algorithm mainly depends on six different steps.

- Selection of some set of points from both point clouds
- Matching of the points between two-point clouds
- Weighting the corresponding pair
- Rejecting specific pairs based on checking all the pairs
- Assign an error metric
- Minimizing the error.

The algorithm is a continuous process of finding minimum distance. So, it is terminated based on any three approaches: (1) Given a threshold on distance measurement, (2) Difference of distance values between two successive iterations, and (3) Reached the maximum iteration.

Non-rigid Registration The transformation happens with respect to the localized stretching of the image. It allows non-uniform mapping between images. AAM is also used for non-rigid registration by using piecewise affine transformation around each landmark [24]. It is also used for small changes in faces like expression. Facial sequence registration by a well-known technique SIFT-flow [25] is also possible. Nowadays, most exciting and challenging work in registration involves the non-rigid-based registration due to soft tissue deformation during imaging.

Affine Transformation Affine transformation is similar to rigid transformation, but it includes extra transformation operations like scaling, shearing. So, after registration, the shape of the object will be changed. Due to the change of shape or scale, it is treated as non-rigid.

4.2 Point-Based Registration

Point-based registration is also termed as point matching. The two individual points' set are aligned through spatial transformation. Finding out the fiducial points correctly is one important part of this type of registration. The active appearance model (AAM) [18] is widely used for point-based registration. This technique is also possible using local feature-based that extracts from the image directly like corner detection of an image. Point set technique can work on raw 3D point cloud data. Let $\{P, Q\}$ be two finite point sets in a finite-dimensional vector space $\mathbb{R}^d$. After spatial transformation, T, the registered point sets $T(P)$. Now the main focus is to minimize the Euclidean distance between two points' sets as in Eq. 2.

$$\text{dist}(T(P), Q) = \sum_{p \in T(P)} \sum_{q \in Q} (p - q)^2 \qquad (2)$$

Rigid and non-rigid registrations are also part of the point-based registration.

4.3   Part-Based Registration

The registration process with respect to some major parts of the object can be classified as parts' registration. According to the human face, the registration in terms of eyes, mouth, etc requires spatial consistency of each part. The parameters—number, size, and location of parts—are varied according to the requirement. Considering AAM—the parts are localized through patches around the detected landmarks. Part-based registration is treated non-rigid/affine registration when shape of object changes.

4.4   Registration Using Model-Based Approach

The model-based registrations are like active appearance model (AAM) [18] and active shape model (ASM) [26]-based alignment. These models are mainly used for human faces.

AAM is one useful method for locating objects. At the time AAM creation, many images with different shapes are used for training the model after that selects some crucial points for annotation—a series of transformations such as PCA, mean shape that used for the face alignment. Initially, estimate the model in any fixed position. Further, some suggestive movements give proper alignment.

ASM is a statistical model of the shape of the object that iteratively fits the example object in a new image. ASM is handled by point distribution model (PDM). It is a local texture model for face alignment.

4.5   Pros and Cons of Registration Techniques

The registration techniques are mainly divided into two parts: rigid and non-rigid. From the transformation point of view, the rotation and translation operations are used for the rigid registration process, whereas the non-rigid registration performs nonlinear transformation such as scaling and shearing. These non-rigid transformation operations are also termed as an affine transformation. The discussion on the advantages and disadvantages of rigid and non-rigid registrations is given below.

The rigid global transformations are easy to implement. For 3D image registration, six parameters are used (three for translation and three for rotation). It is the well-known approach as a research tool, whereas non-rigid registration allows non-uniform mapping between images. The non-rigid registration process corrects small and varying discrepancies by deforming one image to match with the other. However, the non-rigid technique is partly used due to difficulties in a validation. The validation states that a registration algorithm applied to any data of an application consistently succeeds with a maximum (or average) error acceptable for the application. Landmark-based rigid registration validates the system quickly. It is proved that landmark distance between the source and target model is always fixed for rigid registration, so the error analysis has been studied and calculated target registration error for the whole volume, and that can be estimated from the position of landmark accurately. In the case of non-rigid registration, the landmark distances between source and target models always differ. So, the maximum or average registration error can not be calculated from the landmark positions.

5   History of Registration Technique for 3D Face Analysis

3D face registration is one of the essential preprocessing steps for developing any robust 3D face analysis system. Pose alignment cannot be handled easily in the case of the 2D face due to insufficient image depth data, whereas 3D data easily overcome the problems. In this section, we have discussed some of the recently proposed techniques of 3D face registration.

In [27], the authors have developed a 3D face mesh of different orientations from the 2D stereo images. After getting the 3D models of different orientations of any subject, they have considered two symmetric oriented view models for registration. They have identified some significant landmark points on those symmetric view 3D models. Further, they propose a spatial–temporal logical correlation (STLC) algorithm for accurate disparity mapping with a coarse and refined strategy. Finally, 3D data of the whole face region obtained by fusing the left and right measurements. In [28], the registration of pose-variant inputs of this recognition system uses three-dimensional variance of the facial data for transforming it into frontal pose. Unlike the traditional registration algorithm with respect to two images (source and target), the proposed algorithm works with a single image for registration. The test image rotates and calculates variance of XY, YZ, and ZX planes separately. The resulting variance gives the output of registered image. In [29], the authors introduce a novel 3D alignment method that works on frontal face as well as right and left profile face images. At first, detection of pose is done by nose tip-based pose learning approach. After detection, coarse to fine registration is happened by using L_2 norm minimization that is followed by a transformation process to align the whole face. The alignment process is done in three separate planes: *XY, YZ,* and *ZX.* Accurate nose-tip detection in different probe faces and frontal faces is the main challenge of this proposed work. Next, in [30], 3D face sequences are used for the implementation of the proposed registration technique. The authors have used regression-based static registration method that is improved by spatiotemporal modeling to exploit redundancies over space and time. The method is fully geometric-based approach, not require any predetermined landmarks as input. The main innovation is to handle dynamic 3D data for registration, whereas most of the proposed techniques are worked with static data. Another way of 3D face registration is shown in [31]. The 3D morphable model (3DMM) constructed for developing a face recognition system. The authors have considered different types of populations like Black and Chinese. Further, the morphable model was constructed by mixture of Gaussian subpopulations that are named Gaussian mixture 3D morphable model (GM3DMM). The global population treated as mixture of subpopulations. As per previous works, these models are used for 2D face recognition, where the 2D images are mapped onto model. These models have also solved the pose variation issue of face image. Next, in [32], the authors have constructed three-dimensional facial structure from single image using an unsupervised learning approach. Further, remove pose variation problem by doing viewpoint transformation on the 3D structure. The affine transformation warps the source to target face using neural network. The authors have given the name of the proposed network as Depth Estimation-Pose Transformation Hybrid Networks (DepthNet in short). In the case of rotation, the predicted depth value and the affine parameters are generated through pseudo-inverse transformation of some key points of source and target. In [33], 3D face registration was done by using a novel conformal mapping algorithm. The mapping algorithm happens through harmonic energy. Minimizing the harmonic energy by a specific boundary condition of the surface and obtain conformal mapping. Facial surface matching is done by conformal mapping. The method is robust to complicated topologies and not sensitive to quality of surface. In [34], the improved ICP algorithm was used for accurate registration. The registration process of this system divided into two stages. At the first stage, rough registration has been done on depth maps. Initially, selecting some SIFT feature points on the depth map image. The transformation was done for the correct matching of feature points. Here, SVD technique was used for transformation. After that accurate registration is done through ICP algorithm on 3D point clouds. In [35], the authors have introduced a method that solves the problem of face alignment and 3D face reconstruction simultaneously from 2D face images with pose and expression variations. Two sets regressors are used for landmarks detection and 3D shape detection. There is a mapping from 3D-to-2D matrix. The landmark regressor is used to detect landmarks on 2D face images. The shape regressor is used to generate 3D shapes. The regressors adjusted 3D face shape and landmark points. From this adjustment, finally it produced the pose–expression–normalized 3D (PEN3D) face shapes. In [36], the authors wanted to focus the ICP technique in non-rigid form. They find out different non-rigid portions of the face and combine those non-rigid ICP algorithms. The main contribution is describing and fitting 3D faces in new form by learning a local statistical model for facial parts, and ICP algorithm is applied to individual parts before combining to a single model. In [37], the authors have developed a 3D morphable model. This type of model has some drawbacks. It has taken huge numbers of good quality face scans for training the model. For removing the drawback, the scans need to be brought into dense correspondence with a mesh registration algorithm. Similar to previous work, in [38], the 3D morphable face model (3DMM) from 2D image was used for removing pose and illumination variations related issues. The extended idea of this proposed model remaps the pose-variant faces into frontal face. The process is done by fusion of 2D pose-variant textured image and 3D morphable model. In [39], the registration process is done in two stages. In the coarse registration stage, the principle components of unregistered and frontal face images were used for finding the smallest correspondence between each point on the three-dimensional surface. The centroid of the two-point clouds is used to calculate the smallest distance. Next, fine registration is computed using ICP technique. In [40], the authors have worked with 3D facial motion sequence that analyzes statistically the change of 3D face shape in motion. They have introduced landmark detection on the 3D motion sequence-based Markov random fields. To compute the registration, multi-linear model is used. They have used a learned multi-linear model for fully automatic registration of motion sequences of 3D faces. The model previously learned through using a registered 3D face database. In [41], a mathematical model for 3D face registration is used with pose varying across $0°-90°$ in both +ve and −ve orientations. Initially, choose a basic landmark model against whom the images are to be registered. Hausdorff distance metric is used to calculate the distance for alignment. In this technique, all types of pose variations, like pitch, yaw, and roll, were used. In [42], first, they have introduced a feature localization algorithm based on binary neural network technique—k-nearest neighbor advanced uncertain reasoning architecture (kNN AURA) to train and find out the nose tip. The face area has been detected based on the nose tip. After that, registration of 3D faces is done through some integrated steps. First, PCA is used for removing misalignment roughly and then minimizes the misalignment about oy- and oz-axes using symmetry of the faces. Finally, ICP algorithm was used for correct alignment on the expression invariant areas. In [43], the nose tip has been used for 3D face registration using maximum intensity algorithm. The rotational and translational parameters are used for registration. The deformation angle of pose-variant faces calculated based on the maximum intensity point that denotes nose-tip point. The method overcomes the problem of ICP, Procrustes analysis-based registration. Before applying ICP, there should be rough alignment first. In Procrustes analysis, various landmarks must be detected first. The proposed method does not work with high pose. In [44], a non-rigid registration algorithm is used that deforms a 3D template mesh model to the best match depth image. The authors have not used any landmark detection technique. They introduced image-based non-rigid deformation with an automatic feature detection process. In [45], the authors introduced topology adaptive non-rigid methods for registering 3D facial scans. It combines non-rigid mechanisms and mesh subdivisions to improve the drawbacks of previous traditional methods. In [46], the authors initially localized the nose tip on the range image in an efficient way. Further,

Table 2 Importance of registration based on 3D face recognition results of three individual 3D face databases

Database used	PCA (100 components)		PCA (150 components)	
	Without ICP	With ICP	Without ICP	With ICP
Frav3D	56.31	69.52	56.79	70
GavabDB	31.39	50.73	32.12	51.09
Casia3D	64.27	68.02	64.92	68.16

calculate angular deviation in between nose tip of frontal face and pose-variant faces with respect to the difference between x- and y-positions of the depth image. After

6 Result Analysis and Discussion

In this section, first, we have discussed the actual need of registration process in any face analysis system. We have experimentally analyzed the issue by developing a simple 3D face recognition system on three well-known 3D face databases: Frav3D, GavabDB, and Casia3D. We have used a simple PCA technique for feature reduction followed by the support vector machine (SVM) classifier for classification. All three input databases consist of different varieties of face data including pose variation. First, we have registered all the pose-variant faces using the well-known ICP algorithm, where 3D point cloud is considered as input. Further, we have constructed depth images from 3D point clouds for the input of recognition system. Two sets of components, 100 and 150 components from PCA technique, are used for classification. The twofold cross-validation applied for training and testing the recognition system on individual datasets including neutral and other variations of faces. Table 2, given below, illustrates comparative analysis of face recognition accuracies between inputs with ICP and inputs without ICP.

From the result set of the above table, it is clear that registration is a primary preprocessing technique in any face analysis system. Now, we have discussed the experimental results of previous different registration techniques in Table 3; those techniques are already discussed in Sect. 5.

7 Conclusion

Face image registration is one of the necessary processes in any facial analysis system for maintaining the robustness of system. In this chapter, we have emphasized the concept of registration, along with some current works on pose detection. It has shown that the importance of face registration with respect to face recognition is enormous.

Table 3 Discussion on experimental result on different proposed registration techniques

Proposed papers	Registration type	Input for registration	Database used	Registration result
Zhou et al. [27]	Rigid	Real-time stereo-based 3D face data	Real-time data	Considering maximum error, mean error, and standard deviation error for this experiment. When the number of stereo image pairs is $N = 4$, the measurement accuracy in terms of max error $= 0.727$, mean error $= 0.020$, standard deviation error $= 0.035$. Considered $N = 1$, then max error $= 1.511$, mean error $= 0.030$, standard deviation error $= 0.048$
Ratyal et al. [28]	Rigid	3D point cloud	FRGC v2.0, GavabDB	NA
Ratyal et al. [29]	Rigid	3D point cloud and depth image	GavabDB, Bosphorus3D, UMB-DB, FRGC v2.0	The proposed registration algorithm accurately aligned and minimized L2 norms of 99.82%, 100% (nonoccluded), 100, and 99.95% subjects from GavabDB, Bosphorus3D, UMB-DB, and FRGC v2.0 databases
Abrevaya et al. [30]	Non-rigid	The sequence of 3D face scans	D3DFACS, BU-4DFE, BP4D-spontaneous	95% registration accuracy in BP4D database, 100% success in the other two databases

(continued)

Table 3 (continued)

Proposed papers	Registration type	Input for registration	Database used	Registration result
Koppen et al. [31]	Model fitting	3D face data	Caucasian, Chinese, African 3D face data	NA
Moniz et al. [32]	Non-rigid	Single 2D image	3DFAW	Considering DepthNet, values of average correlation between source and target are 27.36, 26.32, and 22.34 for a left, frontal, and right poses. Similarly, considering DepthNet + GAN, values of average correlation between source and target are 59.69, 58.68, and 59.76 for left, frontal, and right poses
Qian et al. [33]	Non-rigid (point-based)	3D face scan	Nearly, 40,000 faces are used in this experiment	NA
Yang et al. [34]	Rigid	Real-time RFB and depth data captured by Kinect Xbox	Real-time data	NA
Liu et al. [35]	Rigid	3D face model	BU3DFE, AFLW, AFLW2000 3D	Considering mean absolute error (MAE) of reconstruction error on pose-variant images of −90° to +90° of the BU3DFE database. The average MAE for all the pose-variant images of the BU3DFE database is 2.8

(continued)

Table 3 (continued)

Proposed papers	Registration type	Input for registration	Database used	Registration result
Cheng et al. [36]	Non-rigid (part-based)	3D depth face	BU3DFE	NA
Kittler et al. [37]	Rigid	3D morphable face model	CMU-PIE and Multi-PIE database	NA
Huber et al. [38]	Rigid (point-based)	3D morphable model	3D face model	NA
Bagchi et al. [39]	Rigid	3D point clouds	GavabDB	NA
Bolkart et al. [40]	Rigid (point-based)	3D face sequence	BU4DFE as face motion sequence data	The registration accuracy was generated by 93.8% for 470 sequences. The distance comparison is 56% successful when considering distance difference is less than 1 mm, and the median error of per-vertex is lower than 1 mm for 73% vertices
Bagchi et al. [41]	Rigid (point-based)	3D point cloud	Bosphorus3D	Calculating rotational error against different sets of the pose. The average rotational error was ranging from 0.003 to 0.009, with pose ranging from 0° to 20° and from 40° to 45°. For poses with 90°, the average rotational error was 0.004
Ju et al. [42]	Rigid	3D point cloud	FRGC v2.0	Calculating mean square error (MSE) for non-neutral face is 0.2550 mm and for neutral face is 0.1940 mm

(continued)

Table 3 (continued)

Proposed papers	Registration type	Input for registration	Database used	Registration result
Bagchi et al. [43]	Rigid (point-based)	3D point cloud	FRAV3D	The success rate of registration across the X-axis: 79.66%, across the Y-axis: 79.66%, and across Z-axis: 67.7%. The overall performance rate of registration is 75.84%
Chen et al. [44]	Non-rigid	RGBD face image	Real-time 3D data recorded by Kinect	For quantitative evaluation, the reconstruction error is 1.2112 mm in X (3D space), 1.5817 mm in Y (3D space), and 2.1866 mm in Z (3D space)
Gao et al. [45]	Non-rigid	3D face	3D face scanned by the structured light scanner	Considering three different 3D scans. The registration error of Scan1 is 0.24%, Scan2 is 0.18%, and Scan3 is 0.29%
Bagchi et al. [46]	Rigid (point-based)	3D face	FRAV3D	NA

NA Not Available

Over the past decades, there exist various registration techniques that are used in different applications. Here, we have mainly focused on the 3D face registration technique that followed by landmark detection, face recognition, etc. There are two groups of registration techniques: rigid and non-rigid. The iterative closest point (ICP) is one of the popular rigid registration techniques. Other than ICP, various rigid- and non-rigid-based techniques are existed. The advantages and disadvantages of different techniques followed by experimental analysis with respect to 3D faces are discussed in this chapter. It can be concluded that the chapter will provide the details of all the recent registration techniques, which are worked on either 3D face point cloud or 3D face depth images. In the future, the registration process can be possible using 3D voxel representation as input.

Acknowledgements The first author is grateful to the Ministry of Electronics and Information Technology (MeitY), Government of India, for the grant of the Visvesvaraya Doctoral Fellowship Award. The authors are also thankful to CMATER laboratory of the Department of Computer Science and Engineering, Jadavpur University, Kolkata, India, for providing the necessary infrastructure for this work.

References

1. Srivastava, A., Mane, S., Shah, A., Shrivastava, N., & Thakare, B. (2017). A survey of face detection algorithms. In *International Conference on Inventive Systems and Control* (ICISC).
2. Abate, A. F., Nappi, M., Riccio, D., & Sabatino, G. (2007). 2D and 3D face recognition: A survey. *Pattern Recognition Letters, 28*(14), 1885–1906.
3. Tian, Y. L., Kanade, T., & Cohn, J. F. *Facial expression analysis.* Available at: https://www.cs.cmu.edu/~cga/behavior/FEA-Bookchapter.pdf.
4. Lee, E. J., & Kwon K. R. (2007). Automatic face analysis system based on face recognition and facial physiognomy. In M. S. Szczuka (Eds.), *Advances in hybrid information technology. Lecture notes in computer science* (Vol. 4413, pp. 128–138). Springer.
5. Kasthuri, A., & Suruliandi, A. (2017). A survey on face annotation techniques. In *4th International Conference on Advanced Computing and Communication Systems (ICACCS)*.
6. Pietrzyk, U., Herholz, K., Schuster, A., Stockhausen, H. M. V., Lucht, H., & Heiss, W. D. (1996). Clinical applications of registration and fusion of multimodality brain images from PET, SPET, CT and NRI. *European Journal of Radiology, 21*(3), 174–182.
7. Oh, S., & Kim, S. (2017). Deformable image registration in radiation therapy. *Radiation Oncology Journal, 35*(2), 101–111.
8. Simon, A., Castelli, J., Lafond, C., Acosta, O., Haigron, P., Cazoulat, G., & Crevoisier, R. D. (2019). Deformable image registration for radiation therapy: principle, methods, applications and evaluation. *Acta Oncologica, 58*(9), 1225–1237.
9. Boldea, V., Sharp, G., Jiang, S. B., & Sarrut, D. (2008). 4D-CT lung motion estimation with deformable registration: Quantification of motion nonlinearity and hysteresis. *Medical Physics, 33*(3), 1008–1018.
10. Meyer, G. P., Gupta, S., Frosio, I., Reddy, D., & Kautz, J. (2015). Robust model-based 3D head pose estimation. In *IEEE International Conference on Computer Vision (ICCV)* (pp. 3649–3657).
11. Tan, D. J., Tombari, F., & Navab, N. (2015). A combined generalized and subject-specific 3D head pose estimation. In *International Conference on 3D Vision*.

12. Borghi, G., Fabbri, M., Vezzani, R., Calderara, S., & Cucchiara, R. (2017). Face-from-depth for head pose estimation on depth images. *IEEE Transaction on Pattern Analysis and Machine Intelligence.*
13. Bagchi, P., Bhattacharjee, D., Nasipuri, M., & Basu, D. K. (2012). A novel approach in detecting pose orientation of a 3D face required for face registration. In *Proceedings of 47th Annual National Convention of CSI & 1st International Conference on Intelligent Infrastructure.*
14. Li, D., & Pedrycz, W. (2014). A central profile-based 3D face pose estimation. *Pattern Recognition, 47,* 523–534.
15. Ashburner, J., & Friston, K. J. *Rigid body registration.* Available at: https://www.fil.ion.ucl.ac.uk/spm/doc/books/hbf2/pdfs/Ch2.pdf.
16. Jiang, B., Valstar, M., Martinez, B., & Pantic, M. (2014). Dynamic appearance descriptor approach to facial actions temporal modelling. *IEEE Transaction on System, Man, Cybernetics, 44*(2), 161–174.
17. Littleword, G. C., Bartlett, M. S., & Lee, K. (2009). Automatic coding of facial expressions displayed during posed and genuine pain. *Image Vision Computing, 27*(12), 1797–1803.
18. Cootes, T. F., Edwards, G. J., & Taylor, C. J. (2001). Active appearance models. *IEEE Transaction on Pattern Analysis Machine Intelligence, 23*(6), 681–685.
19. Besl, P. J. (1992). A method for registration of 3D shapes. *IEEE Transaction Pattern Analysis and Machine Intelligence, 14*(2), 239–256.
20. Zhang, Z. (1994). Iterative point matching for registration of free-form curves and surfaces. *International Journal of Computer Vision, 13*(2), 119–152.
21. Menq, C. H., Yau, H. T., & Lai, G. Y. (1992). Automated precision measurement of surface profile in CAD-directed inspection. *IEEE Transaction on Robotics and Automation, 8*(2), 268–278.
22. Danielsson, P. E. (1980). Euclidean distance mapping. *Computer Graphics and Image Processing., 14*(3), 227–248.
23. Gruen, A., & Akca D. *Least squares 3D surface matching.* Available at: https://pdfs.semanticscholar.org/790f/ccfa5d36ca0f04de24374faed273af8b5e90.pdf.
24. Lucey, P., Cohn, J., Prkachin, K. M., Solomon, P. E., Chew, S., & Matthews, I. (2012). Painful monitoring: Automatic pain monitoring using the UNBC-McMaster shoulder pain expression archive database. *Image Vision Computing, 30*(3), 197–205.
25. Liu, C., Yuen, J., & Torralba, A. (2011). SIFT flow: Dense correspondence across scenes and its applications. *IEEE Transaction on Pattern Analysis Machine Intelligence, 33*(5), 978–994.
26. Lu, H., & Yang, F. (2014). Active shape model and its application to face alignment. In Y. C. Chen & L. Jain (Eds.), *Subspace methods for pattern recognition in intelligent environment. Studies in computational intelligence* (Vol. 552). Springer.
27. Zou, P., Zhu, J., & You, Z. (2019). 3-D face registration solution with speckle encoding based spatial-temporal logical correlation algorithm. *Optical Express, 27*(15), 21004–21019.
28. Ratyal, N. I., Taj, I. A., Sajid, M., Ali, N., Mahmood, A., & Razzaq, S. (2019). Three-dimensional face recognition using variance-based registration and subject-specific descriptors. *International of Advanced Robotic Systems, 16*(3), 1–16.
29. Ratyal, N. I., Taj, I. A., Sajid, M., Mahmood, A., Razzaq, S., Dar S. H., et al. (2019). Deeply learned pose invariant image analysis with applications in 3D face recognition. *Mathematical Problems in Engineering, 2019*(Article 3547416), 1–21.
30. Abrevaya, V. F., Wuhrer, S., & Boyer, E. (2018). Spatiotemporal modeling for efficient registration of dynamic 3D faces. In *International Conference on 3D Vision.*
31. Koppen, P., Feng, Z. H., Kittler, J., Awais, M., Christmas, W., Wu, X. J., et al. (2018). Gaussian mixture 3D morphable face model. *Pattern Recognition, 74,* 617–628.
32. Moniz, J. R. A., Beckham, C., Rajotte, S., Honari, S., Pal, C. (2018). Unsupervised depth estimation, 3D face rotation and replacement. In *32nd Conference on Neural Information Processing Systems (NeurIPS 2018), Montreal, Canada.*
33. Qian, K., Su K., Zhang, J., & Li, Y. (2018). A 3D face registration algorithm based on conformal mapping. *Concurrency and Computation: Practice and Experience, 30,* e4654. https://doi.org/10.1002/cpe.4654

34. Yang, G., Zeng, R., Dong, A., Yan, X., Tan, Z., & Liu, Y.: Research and application of 3D face modeling algorithm based on ICP accurate alignment. *Journal of Physics Conference Series, 1069*(1).
35. Liu, F., Zhao, Q., Liu, X., & Zeng, D. (2018). Joint face alignment and 3D face reconstruction with application to face recognition. *IEEE Transactions on Pattern Analysis and Machine Intelligence.*
36. Cheng, S., Marras, I., Zafeiriou, S., & Pantic, M. (2017). Statistical non-rigid ICP algorithm and its application to 3D face alignment. *Image and Vision Computing, 58,* 3–12.
37. Kittler, J., Huber, P., Feng, Z. H., Hu, G., & Christmas, W. (2016). 3D morphable face models and their applications. In F. Perales, J. Kittler (Eds.), *Articulated motion and deformable objects. AMDO 2016. Lecture notes in computer science* (Vol. 9756). Cham: Springer.
38. Huber, P., Hu, G., Tena, R., Mortazavian, P., Koppen, W. P., Christmas, W., et al. (2016). A multiresolution 3D morphable face model and fitting framework. In *11th International Joint Conference on Computer Vision, Imaging and Computer Graphics Theory and Applications* (pp. 27–29).
39. Bagchi, P., Bhattacharjee, D., & Nasipuri, M. (2016). Registration of range images using a novel technique of centroid alignment. In R. Chaki, A. Cortesi, K. Saeed, & N. Chaki (Eds.), *Advance computing and systems for security. Advances in intelligent systems and computing* (Vol. 396, pp. 81–89).
40. Bolkart, T., & Wuhrer, S. (2015). 3D faces in motion: fully automatic registration and statistical analysis. *Computer Vision and Image Understanding, 131,* 100–115.
41. Bagchi, P., Bhattacharjee, D., Nasipuri, M.: 3D face recognition across pose extremities. In M. Kumar Kundu, D. Mohapatra, A. Konar, & A. Chakraborty (Eds.), *Advanced computing networking and informatics. Smart innovation, systems and technologies* (Vol. 27, pp. 291–299). Springer
42. Ju, Q. (2013). Robust binary neural networks based 3D face detection and accurate face registration. *International Journal of Computational Intelligence Systems, 6*(4), 669–683.
43. Bagchi, P., Bhattacharjee, D., Nasipuri, M., & Basu, D. K. (2013). A method for nose-tip based 3D face registration using maximum intensity algorithm. In *Proceedings of International Conference of Computation and Communication Advancement.*
44. Chen, Y. L., Wu, H. T., Shi, F., Tong, X., & Chai, J. (2013). Accurate and robust 3D facial capture using a single RGBD camera. In *International Conference on Computer Vision* (pp. 3615–3622).
45. Gao, Y., Li, S., Hao, A., & Zhao, Q. (2012). Topology-adaptive non-rigid registration for 3D facial scans. *International Journal of Future Computer and Communication, 1*(3), 292–295.
46. Bagchi, P., Bhattacharjee, D., Nasipuri, M., & Basu, D. K. (2012). A novel approach for registration of 3D face Images. In *International Conference On Advances In Engineering, Science And Management.*

Image Denoising Using Generative Adversarial Network

Ratnadeep Dey, Debotosh Bhattacharjee, and Mita Nasipuri

Abstract Image denoising is one of the most important and fundamental research areas in the digital image-processing field. A noisy image can mislead image processing-based research. Therefore, image denoising is a critical area of research. In the recent advancement of computer vision, deep learning becomes most powerful tool. Deep learning is solving most of the problems, usually, which were earlier solved by various conventional techniques. The progress of deep learning encourages researchers to apply deep learning-based methods into image denoising also. In recent years, generative adversarial network (GAN) becomes a new avenue in computer vision research. The GANs are adversarial networks with generative capability, and the network has a very vast area of applications. In this chapter, we concentrate on a specific area of application of GAN—image denoising. At first, the traditional denoising techniques are highlighted. Then, we state the underlying architecture of GAN and its modifications. Then, we discuss the way GANs are applied in the area of image denoising. We survey all recent works of GANs in image denoising and categories those work according to the type of input images. In the end, we propose some research directions in this area. The compilation and discussions presented in this chapter regarding image denoising using GAN are new inclusion, and similar survey work is not available for the community.

Keywords Generative adversarial network (GAN) · Image denoising · Deep learning · Image derain

R. Dey (✉) · D. Bhattacharjee · M. Nasipuri
Department of Computer Science and Engineering, Jadavpur University, Kolkata, India
e-mail: ratnadipdey@gmail.com

D. Bhattacharjee
e-mail: debotoshb@hotmail.com

M. Nasipuri
e-mail: mitanasipuri@gmail.com

© Springer Nature Singapore Pte Ltd. 2020
J. K. Mandal and S. Banerjee (eds.), *Intelligent Computing: Image Processing Based Applications*, Advances in Intelligent Systems and Computing 1157,
https://doi.org/10.1007/978-981-15-4288-6_5

1 Introduction

In recent years, the number of per unit areas in image sensors of input devices increases, and for that, cameras become more sensitive to noise. The presence of bit of noise in scenes makes the captured image noisy. Suppose, X and Y are image coordinates in 2D space and a pixel is denoted by (x, y). The noise-free actual image is denoted by $T(x, y)$ and the noisy image is denoted by $D(x, y)$. Then, the exact image and the noisy image are related, as shown in Eq. 1. Where $N(x, y)$ is the additive noise. The noise N follows some distribution.

The noises can be occurred in two ways—(i) due to issues relating to the acquisition system or (ii) by the external instinct. Defocusing of camera is prevalent cause in image denoising in the first category. Other than that different noises can occur in different acquisition systems. The X-ray images are encountered with Poisson noise [1]. In the case of ultrasound images, speckle noise [2] can be observed. In the second category, noise can occur due to external factors. That means, the acquisition system captured degraded scenes. The degradation may be caused by weather degradation or other external instincts. Image denoising is the process of removing those noises from degraded image to recover the actual image. According to Eq. 1, image denoising techniques try to minimize $N(x, y)$ to produce $D(x, y) \approx T(x, y)$.

$$D(x, y) = T(x, y) + N(x, y) \quad x \in X, y \in Y \tag{1}$$

In the field of image denoising, the researcher tries to restore input images and ensures noise-free input images for further analysis.

The image-denoising algorithms are broadly classified into three categories [3]. They are (i) Spatial domain (ii) Transform domain (iii) Learning-based approach.

(i) Spatial domain approach—In this approach, spatial correlation among image pixels comes into play. This type of approach further classified into two subclasses as local and nonlocal filters. In the case of local filters, a correlation among the neighboring pixels is calculated. Gaussian filter [4], Wiener filter [5], least mean squares filter (LMSF) [6], anisotropic filtering [7], trained filter (TF) [8], steering kernel regression (SKR) [9], bilateral filter [10], kernel-based image denoising employing semi-parametric regularization (KSPR) [11], metric Q for tuning parameters in SKR (MSKR) [12] come into local filter subcategory. These types of filters are advantageous in terms of time complexity. However, they are not useful in case of highly noised image. Nonlocal filtering concentrates on similar pixels without restriction of distance. Nonlocal mean (NLM) [13] is the example of nonlocal filtering. Researchers improve [14–16] the NLM filtering as well. The nonlocal filters are good for high-noise conditions, but they suffer over-smoothing problems [17].

(ii) Transform domain approach—Images are transformed into other domains depending on the orthonormal basis. Then, the coefficients of those domains are adjusted to reduce noise. It has been seen that smaller coefficients are the higher frequency part of the image, and these coefficients are highly responsible for image details and noises. Therefore, in this approach, small coefficients are to be manipulated to reduce noise. In this approach, wavelet-based methods [18–21] are prevalent.

(iii) Learning-based approach—This approach is based on the sparse representation of images. A large number of image patches are used to learn the representation, and according to the learning, the images are restored. K-clustering with singular value decomposition (K-SVD) [22], clustering-based sparse representation (CSR) [23], and learned simultaneous sparse coding (LSSR) [24] are some research works in this approach.

In the advancement of deep learning, image denoising is also done by deep learning-based methods. The deep learning-based methods mainly come into the category of learning-based approach. In very recent times, generative adversarial networks [25] become very popular. The GAN mainly falls into the category of deep multimodal representation learning [26]. The GAN is used to generate new data from scratch. However, the GANs are used in the domain of denoising. In this chapter, we discuss the basic architecture of GAN and how it is used in image denoising. Then, we categorize the work in this area according to the type of image data used in the research works. This chapter mainly provides a survey work on the application of GAN in the field of image denoising and according to best of our knowledge, no survey work in this field has been done yet.

The chapter is organized as follows—Sect. 2 describes the underlying architecture of GAN and its modifications; in Sect. 3 categorization of research works in the field of image denoising using GAN is shown and they are discussed elaborately in subsections; Sect. 4 shows some research directions in this domain, and Sect. 5 concludes the chapter.

2 Underlying Architecture of Generative Adversarial Network (GAN)

Goodfellow et al. [25] proposed a new destination in machine learning. The main goal of the network is to generate new data from scratch. Random noise can be the input of the network that generates new data similar to the target dataset. In the work [25], data similar to MNIST [27] database was produced from a random noise input using GAN in its first application. The GAN consists of two networks—Generator and Discriminator. The two competitive networks are working on the principle of game theory. The generator network generates new data, and the discriminator network evaluates the data generated by the generator network. Figure 1 illustrates the basic working principle of the GAN. This figure is used here to describe the GAN.

In Fig. 1, the green block denoted as G represents the generative model, and the blue block denoted as D represents the discriminator block. The block G takes random noise X as input and produces a pattern $G(X)$. The block D trained by a real data Y works as a binary classifier. It classifies between $G(X)$ and $D(Y)$, and the classification result monitors the G to generator new data alike Y. At first, the

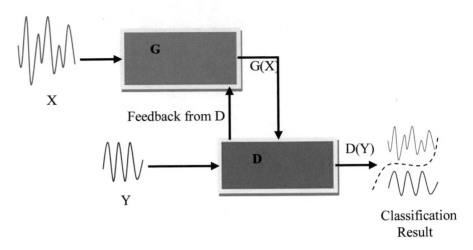

Fig. 1 Basic architecture of GAN

D block is trained with real data or target data (Y). The D block tries to learn the pattern of data Y using convolution operation where the G block generates data using deconvolution operation. The loss function $J^{(D)}$ of discriminator [28] is defined in Eq. 2.

$$J^{(D)} = -\frac{1}{2}\mathbb{E}_{Y \sim p\text{data}} \log D(Y) - \frac{1}{2}\mathbb{E}_X \log(1 - D(G(X))) \tag{2}$$

where $\mathbb{E}$ is the expectation, $D(Y)$ indicates the probability that the block D correctly recognize real data Y, $D(G(X))$ is the probability of determining that the block D correctly identify $G(X)$. The D block mainly works as a classifier. Therefore, the main goal of the D block is to identify the source of the data correctly. As a result, the D block tries to minimize $D(G(X))$. However, the G block tries to generate data similar to Y. That means the goal of G to generate $G(X) \sim Y$. That creates a conflict between the D block and G block. The G block tries to maximize $D(G(X))$ just opposite of the D. Therefore, the loss function $J^{(G)}$ of the block G can be written as Eq. 3.

$$J^{(G)} = -J^{(D)} \tag{3}$$

In the literature, the primary GAN network has been modified for better performance. The modified GAN networks perform better than the primary GAN network. Researchers changed some parts of the underlying network to make it more robust. Some researchers changed the architecture of the network, some work restricted the input signal of the network, and many researchers used different objective functions for better performance. The modifications of GAN network are enlisted in Table 1. In the table, modified networks are described briefly. The name of the networks, type

Table 1 Types of GAN network

Sl. No	Name of the network	Type of modification	Characteristics of the network
1	Deep convolutional generative adversarial networks (DCGAN) [29]	Structural	In the original GAN network, multi-layer perceptron (MLP) model is used. In DCGAN, the MLP is replaced by convolutional neural network (CNN). Deconvolution layers replace the fully connected layer used in generator module in original GAN. This modification accelerates performance
2	Conditional generative adversarial networks (CGANs) [30]	Conditional input	In this modification, input signals are restricted. A conditional variable is introduced here to check the condition of the input variable
3	Adversarial autoencoder (AAE) [31]	Structural	This is the introduction of the autoencoder in the adversarial network
4	Bidirectional generative adversarial networks (BiGAN) [32]	Structural	An encoder is added to GAN to assure the quality of the generated data
5	Adversarially learned inference (ALI) [33]	Structural	Introduce an encoder to learn the distribution of latent features
6	Adversarial generator-encoder network (AGE) [34]	Structural	In this model, the discriminator module is ignored. The model generates data with the help of the generator and the encoder
7	Unrolled generative adversarial network [35]	Objective function	Gradient-based loss function is used in the generator for data generation
8	Wasserstein generative adversarial networks (WGAN) [36]	Objective function	Propose a new metric to address unstable training based on earth-mover distance calculation
9	WGAN-GP [37]	Objective function	Modification of WGAN
10	WGAN-LP [38]	Objective function	Modification of WGAN-GP
11	Energy-based generative adversarial network (EBGAN) [39]	Objective function	Modifies the discriminator as an energy function

of modifications, and unique characteristics of the network are given in the table. Readers can get an overview of the modified GAN networks by going through this table. In the table, we added all existing modifications on GAN networks for generating new data from random noise as per best of our knowledge. According to Table 1, we can conclude that the modifications of essential GAN networks are restricted into three types—(i) Structural modification, (ii) Conditional Input, (iii) Objective function. In structural modification, the network architecture has been modified. In case of conditional modification, some restriction has been put on the input of the network. Modification in objective function comes in the third category.

3   GAN-Based Image Denoising

In the previous section, we have summarized the underlying architecture of GAN and the modifications done on the basic GAN model. The main goal of the GAN network is to generate new data from a random signal. However, the GAN network has a vast application. The GAN successfully applied to computer vision, medical image processing, natural language processing, image denoising, etc. Among those applications, image denoising is a very new application area of GAN. The GAN network can easily be applied to image denoising. To date, as per our knowledge, no survey work has been done on GAN-based image denoising. In this chapter, we discuss image denoising using GAN and its applications. At first, we will discuss the basic working strategy of denoising using the GAN. Then, we will classify the image-denoising applications of the GAN and describe elaborately about the applications. We classify the applications according to the input image used to denoise.

3.1   Basic Strategy of Image Denoising Using GAN

All works of image denoising using the GAN uses nearly the same principle. According to the principle, the discriminator network is trained using a real image. The noisy data is feed into the generator network. As compared to the basic GAN model, the generator network takes the noisy image as input instead of random noise. The discriminator model learns data distribution using real data. The generator model tries to denoise the noisy data according to the learned data distribution in the discriminator. In Fig. 2, the basic working strategy of GAN-based image denoising is shown. Comparing with Fig. 1, the input of the generator block is replaced by noisy images instead of random noise. The input of discriminator block is real or noise-free data. According to this basic working strategy, the researchers contribute their way of solving the problem of image denoising. They provide different modifications according to the input image. Some researchers propose a general network to denoise any noisy image. In the next section, we will discuss the research works done till now on image denoising using GAN.

Noisy Image

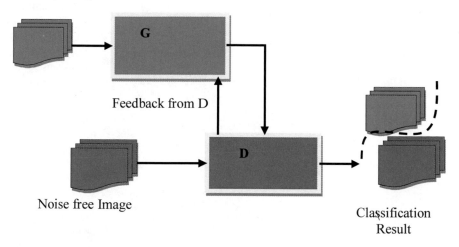

Fig. 2 Working principle of image denoising using GAN

3.2 Research Works on Image Denoising Using GAN

The research works until now can be classified into two categories—(i) Domain-specific and (ii) Domain-independent researches. In the domain-specific type, researchers concentrate on a specific type of image. They proposed GAN-based model to denoise particular types of images. In the domain-independent category, the researchers propose general GAN-based image-denoising models, which can be used to denoise any noisy images.

The domain-specific category can further be divided according to the domain type. That means the category is broken in subcategories according to the type of input images used for denoising. The domain-specific category is divided into five subcategories—(i) Natural image degraded by environmental stimuli (ii) Medical imaging (iii) micro-doppler imaging (iv) SAR (v) Space-based Imaging.

The independent domain category cannot be divided further according to the input image used because these proposed models can be worked on any image. Therefore, independent domain category is not divided further.

Table 2 enlists the categories and subcategories of research works on image denoising using GAN.

Now, we discuss each category of image-denoising technique in detail. At first, the domain-specific category is discussed along with its subcategories. Then, the independent domain category is discussed.

Table 2 Classification of GAN-based image-denoising networks

Sl. No	Category	Subcategory	Specification	Research works
1	Domain specific	Natural image degraded by environmental stimuli	Environmental stimuli like rain, fog, haze degraded images. In this type of work, researchers try to remove noises and recover images for further processing	ACA-net [40], CANDY [41], DCPDN [42], DHGAN [43], GAN based ALPR system [44], ID-CGAN [45], snow removal using GAN [46], reflection removal using GAN [47]
		Medical imaging	Research works concentrating on medical image denoising	DAGAN [48], GAN based OCT image denoising [49], CT image denoising using GAN [50], motion-blind blur removal for CT Images using GAN [51], DRNN [52], noise removal from low-dose X-ray CT images [53], visual attention network for LDCT denoising [54]
		Micro-doppler imaging	This is radar sensor-based applications. In this work, the micro-doppler spectrogram is denoised	Micro-doppler spectrogram denoising using GAN [55]
		SAR	Synthetic aperture radar (SAR) images are the target domain here. This type of images is affected by speckle noise	Denoising synthetic aperture radar (SAR) image [56]

(continued)

Table 2 (continued)

Sl. No	Category	Subcategory	Specification	Research works
		Space-based imaging	Images of space are the central area of concentration	Denoising of space-based image using GAN [57]
2	Domain independent	–	This type of denoising network can denoise any image. Input images are not specific	RCGAN [58], RDS-GAN [59], denoising using CGAN [60], single image denoising and super-resolution [61], denoising using GAN [62], blind denoising [63], DAA [64], Laplacian pyramid based GAN, texture preserving GAN [65]

3.2.1   Domain-Specific Denoising Techniques Using GAN

In this category of research, researchers implement denoising algorithms for a specific type of image. According to the type of images, the category is subdivided into several subcategories. Research works on these categories are discussed elaborately in following discussions.

Natural Image Degraded by Environmental Stimuli

In this subcategory, the researchers try to denoise the images degraded by environmental stimuli. Images can be degraded by different weather conditions like rain, snow, fog, poor illumination, etc., in time of the accusation. The researchers try to remove those noises caused by environmental stimuli by applying different types of GAN networks. Four types of research works can be found in the literature. They are as follows.

(i)   **Haze removal**—Fog, dust, and mist are environment stimuli, which cause blurring and haze in images. The GAN network has been applied successfully to remove haze from input images. Swami et al. proposed a network named CANDY [41] to remove haze from input image. The authors claimed that they were the first to apply GAN in case of removing haze from input image. They used six convolution layers, followed by six deconvolution layers and a Tanh output layer to prepare the generator model. Batch normalization and PReLU activation layers separate each convolution layer. The batch-normalization layers and the ReLU activation layers separate the deconvolution layers. The discriminator network consists of seven convolution networks with LeakyReLU activation layers. The CANDY network was evaluated on a synthetically image database created on their own and got promising results. DHGAN [43] is another work addressing the same problem. This work is inspired by CGAN net. The generator of DHGAN is trained using the patches of scale-reduced input images instead of standard input image patches. This modification helps the generator to capture more features than regular training. The discriminator is trained by most considerable binary cross-entropy loss. The model was evaluated on NTIRE2018 Image Dehazing Challenge dataset and ranked second in the competition. In DCPDN [42] model introduces a concept of joint discriminator, and the generator part consists of three module—dense pyramid for haze estimation, atmospheric light estimation, and dehazing network. In the discriminator, a new loss function called edge-preserving loss function was used. The model was evaluated by two synthetic and one real-world datasets. In an application-based work [44], authors tried to deblur the image of license plates of moving vehicles acquired in hazy conditions. This automatic license plate recognition-based application used the GAN model for correctly identifying license plates.

(ii) **Low light image Enhancement**—The work ACA-net [40] comes under this class. The network uses the characteristics of the GAN to remove problems caused by traditional CNN-based models. The traditional CNN-based models suffer from the presence of visual artifact in output image. The ACA-net obtained good results compared to state-of-the-art techniques. In this work, multi-scale context aggregation network (CAN) [66] was used as the generator. The generator was trained by the concatenation of two enhanced images by two different gamma functions and the input image. The generator used L1 pixel-wise reconstruction loss function. The discriminator used adversarial loss.

(iii) **Rain removal**—The ID-CGAN [45] network was proposed to remove rain strikes from images using the concept of adversarial network. The model used the conditional GAN [30] model as a base network. The authors innovated a new loss function, especially for draining.

(iv) **Snow Removal**—Li et al. [46] proposed a GAN-based model to remove snowflakes from images. The model was inspired by the CGAN [30] model. In this work, the generator consists of two modules—(a) Clean background model, (b) Snow mask estimation model. The clean background model used to generate clean images where the snow mask estimation model estimated snow models. The work was evaluated on the database Snow 100 K^2.

(v) **Image-reflection removal**—The work [47] was proposed to remove reflections from input images using the concepts of the adversarial net in two stages. In the first stage, the authors proposed a loss function to suppress the reflections. In the second stage, the background was recovered. The entire result was inserted into the GAN for reconstructing reflection-free images.

Medical Imaging

The research work in the field of medical image processing is mainly based on images acquired by different radiological instruments. The representation of such images is different from standard images. There are plenty of types of images like CT images, X-ray images, ultrasonic images, etc. The images are affected by noises caused by acquisition devices. The medical images can be denoised using the GAN. In this section, we discuss the research works done on medical image denoising according to the types of medical images.

(i) **Computed Tomography (CT) images**—Low-dose CT is used in medical imaging to reduce the effect of radiation on the human body. However, the low-dose CT image is noisy and full of artifacts. Yang et al. [50] proposed a denoising technique using novel GAN model WGAN. In the WGAN, the data distribution comparison was made by Wasserstein distance. The work was evaluated on the database of the 2016 NIH-AAPM-Mayo Clinic Low-Dose CT Grand Challenge [67]. Another work [54] also evaluated on the same database. In that work, the GAN network is trained using visual attention knowledge of CT images. This innovation produced better results than the state-of-the-art techniques. Lyu et al. [51] worked on CT image deblurring to remove motion blur by using the WGAN and tested the model on their own created blur-sharp database. Another work [53] was removing the noises from X-ray CT image using the objective function of the f-GAN.

(ii) **Optical coherence tomography (OCT) image**—This type of image suffers from speckle noise. Yu et al. [49] proposed DNGAN to denoise OCT images. In this network model, the generator model consisted of more dense layers that helped in denoising.

(iii) **Compressed sensing MRI (CS-MRI) image**—The CS-MRI imaging technique is more robust than normal MRI images. Yang [48] proposed GAN-based model DAGAN for reconstructing the CS-MRI image. The underlying architecture of the model is CGAN [30]. In this work, the refined learning process was used to stabilize U-net-based generator. They proposed a novel content loss function that was used with adversarial loss for better texture preserving output. They evaluated their work on MICCAI 2013 grand challenge dataset [68].

(iv) **Ultrasound Imaging**—The ultrasound images are affected by speckle noise. In work [52], the researchers tried to denoise ultrasound noise. In this work, a novel model despeckling residual neural network has replaced the generator. The discriminator worked with a new structural loss along with adversarial loss.

Micro-Doppler Imaging

Micro-doppler spectrogram is used to analyze the movement of the target in radio-sensor images. The micro-doppler images can be affected by different noises. Traditional CNN-based denoising algorithms cannot remove the noises. However, D. Huang et al. [55] prove that the GAN can be used to denoise micro-doppler images. They used MOtionCAPture (MOCAP) [69] database for their experiment.

SAR

Synthetic-aperture radar (SAR) images suffer from speckle noise. ID-GAN [56] was proposed to denoise the SAR image. The network was trained end-to-end using three losses—Euclidian, adversarial, and perceptual. The networks used here were encoder–decoder type networks. They tested their network on their own created dataset, and the evolution results outperformed state-of-the-art techniques.

Space-Based Imaging

Chen et al. [57] applied WGAN with perceptual loss and adversarial loss to remove motion blur from space-based image of the NanoSats mission.

3.2.2   Domain-Independent denoising Techniques Using GAN

The GAN-based image-noising techniques, which are not designed to denoise aspecific type of image, are discussed in this section. Therefore, the research works cannot be categorized according to the input image. The works are discussed thoroughly one by one.

Chen et al. [61] contributed first in this research domain. They proposed denoising and super resolution via a generative adversarial network (DSGAN) to use the GAN first time for image denoising. The DSGAN can denoise and adjust the resolution of image at the same time. The generator module was trained with noisy images and lower resolution images both at the same time. Residual block with identical layout with batch-normalization layers and the parametric ReLU as activation function were used in the generator. The discriminator module was trained with ground truth data. They evaluated their proposed model on synthetic data, which was created by their own from ImageNet [70] database. Chen et al. [63] addressed the blind denoising problem. In their work, they used the discriminator block of the GAN to estimate the noise model by training using noisy images. Then, according to the learned model, they generated noise samples. The noise samples were passed into normal convolution network model for denoising the noisy images. They named the model as GAN–CNN-Based Blind Denoiser (GCBD). They used three datasets for their work—BSD68 [71], DND [72], and NIGHT. The recurrent conditional generative adversarial network (RCGAN) [58] introduced scale-recurrent generator to extract Spatio–temporal features to generate sharp image in coarse to fine manners. They used receptive field recurrent discriminator to evaluate local and global performances of the generator. They have used GOPRO [73] dataset for their work. Wu et al. [59] proposed RDS-GAN-stack of denoising networks. Stage I predicts the noise model. Stage II was used as the generator to generate denoised images, and at stage III, the discriminator block evaluates the quality of denoised images generated by the generator. They used the database DIV2K train LR bicubic X4 [74] for their training and the databases BSD68, CBSD68, Kodak24, McMaster [75], Set14 [76], and RNI15 [77] were used for testing. ZhiPing [65] divided the generator module into four parts—(i) feature extraction layer (ii) feature denoising layers (iii) high-level feature extraction layers, and (iv) feature fusion layers. The innovation helped them to achieve decent performance. Creswell [64] introduced autoencoder with the GAN for image denoising. Nuthna et al. [60] used CGAN [30] for image denoising. Alsaiari et al. [62] used the GAN-based denoising techniques in 3D animation field and got promising results.

4 Discussion and Future Research Direction

In the previous section, we have discussed different image-denoising techniques using generative adversarial network present in the literature. The basic principle of the research works is to train the generator module with a noisy image and train the discriminator module with noise-free image for denoising. The researchers contributed in their way by keeping the basic working principle intact. The different concept was used to create the generator and the discriminator module for specific application and better performance. The GAN architecture has a drawback of unstable training process. Researchers applied novel loss functions to get rid of the problem. Overall, the discussed research works ensure that the GAN can be used successfully in the research domain of image denoising.

If we think critically, then, we can identify some limitations of this approach. The GAN-based techniques are entirely data driven. The model learns the data structure of ground truth data or noise model by go through massive data. The amount of data controls the performance of the denoising model. The model performs better if it is trained with a large number of data. However, in real-life applications, it is difficult to get a considerable number of noisy data. Insufficient data may lead to failure. Most of the researchers evaluated their proposed denoising models on synthetic data due to insufficient noisy data. The availability of data is one of the main concerns in the GAN-based denoising techniques. The training time of the denoising model is considerable. This may be a problem if we try to use this system in real-time application problems.

In the future, the research orientation in this domain should be the reduction of data dependency of the model. Combination of traditional machine learning techniques and convolution operations can be used to create the discriminator module. Image regression may be fused with a convolution network to predict appropriate data model efficiently with less amount of data. The generator module also tries to add some traditional image quality enhancing techniques to generate denoised image. These modifications may reduce the data dependency of the denoising techniques. These results lower training time as the amount of data is reduced. Therefore, overall system will produce desired result in less amount of time. We think these modifications will help to denoise images using the GAN more appropriately and efficiently.

5   Conclusion

In this chapter, we show the applicability of the generative adversarial network in the field of image denoising. We have introduced our chapter by discussing different types of traditional image-denoising techniques. Then, we have stated the working principle of the generative adversarial network and highlighted the modifications done on the base model of the GAN. After that, we have reached into the main discussion topic of the chapter—image denoising using the GAN. We have reviewed all research works related to the subject. We categorized them into two categories according to the type of input image used in their research work. The categories are also divided into subcategories. We have discussed every prior research works elaborately according to the categories and subcategories. After the thorough discussion, we have identified some limitations of the prior research works and have stated some future directions in this research domain. According to the best of our knowledge, we have surveyed almost all research works on this topic, and the review work is very new in this area of research. This chapter will be helpful for researchers to get knowledge about the image-denoising techniques and the modifications of the techniques arise according to the time.

References

1. Kumar, N., & Nachamai, M. (2017). Noise removal and filtering techniques used in medical images. *Oriental Journal of Computer Science and Technology, 10*(1), 103–113. ISSN: 0974–6471.
2. Patel, N., Shah, A., Mistry, M., & Dangarwala, K. (2014). A study of digital image filtering techniques in spatial image processing. In *International Conference on Convergence of Technology, IEEE.*
3. Shao, L., Yan, R., Li, X., & Liu, Y. (2014). From heuristic optimization to dictionary learning: A review and comprehensive comparison of image denoising algorithms. *IEEE Transactions on Cybernetics, 44*(7), July 2014. https://doi.org/10.1109/tcyb.2013.2278548.
4. Shapiro, L., & Stockman, G. (2001). *Computer vision.* Englewood Cliffs, NJ, USA: Prentice-Hall.
5. Wiener, N. (1949). *Extrapolation, interpolation, and smoothing of stationary time series.* New York, NY, USA: Wiley.

6. Widrow, B., & Haykin, S. (2003). *Least-mean-square adaptive filters*. New York, NY, USA: Wiley-IEEE.
7. Yang, G. Z., Burger, P., Firmin, D. N., & Underwood, S. R. (1995). Structure adaptive anisotropic filtering. In *Proceedings of IEEE International Conference on Image Process* (pp. 717–721). Edinburgh, U.K.
8. Shao, L., Zhang, H., & de Haan, G. (2008). An overview and performance evaluation of classification-based least-squares trained filters. *IEEE Transactions on Image Processing, 17*(10), 1772–1782.
9. Takeda, H., Farsiu, S., & Milanfar, P. (2007). Kernel regression for image processing and reconstruction. *IEEE Transactions on Image Processing, 16*(2), 349–366.
10. Tomasi, C., & Manduchi, R. (1998). Bilateral filtering for gray and color images. In *Proceedings of 6th International Conference on Computer Vision* (pp. 839–846).
11. Bouboulis, P., Slavakis, K., & Theodoridis, S. (2010). Adaptive kernel-based image denoising employing semi-parametric regularization. *IEEE Transactions on Image Processing, 19*(6), 1465–1479.
12. Zhu, X., & Milanfar, P. (2010). Automatic parameter selection for denoising algorithms using a no-reference measure of image content. *IEEE Transactions on Image Processing, 19*(12), 3116–3132.
13. Buades, A., Coll, B., & Morel, J. M. (2005) A nonlocal algorithm for image denoising. In *Proceedings of IEEE International Conference Computer Vision Pattern Recognition* (Vol. 2, pp. 60–65).
14. Coupe, P., Yger, P., Prima, S., Hellier, P., Kervrann, C., & Barillot, C. (2008). An optimized blockwise nonlocal means denoising filter for 3-D magnetic resonance images. *IEEE Transactions on Medical Imaging, 27*(4), 425–441.
15. Tschumperle, D., & Brun, L. (2009). Non-local image smoothing by applying anisotropic diffusion PDE's in the space of patches. In *Proceedings of IEEE International Conference on Image Process* (pp. 2957–2960).
16. Sven, G., Sebastian, Z., & Joachim, W. (2011). Rotationally invariant similarity measures for nonlocal image denoising. *Journal of Visual Communication and Image Representation, 22*(2), 117–130.
17. Brailean, J. C., Kleihorst, R. P., Efstratiadis, S., Katsaggelos, A. K., & Lagendijk, R. L. (1995). Noise reduction filters for dynamic image sequences: A review. *Proceedings of the IEEE, 83*(9), 1272–1292.
18. Dabov, K., Foi, A., Katkovnik, V., & Egiazarian, K. (2007). Image denoising by sparse 3-D transform-domain collaborative filtering. *IEEE Transactions on Image Processing, 16*(8), 2080–2095.
19. Portilla, J., Strela, V., Wainwright, M. J., & Simoncelli, E. P. (2003). Image denoising using scale mixtures of Gaussians in the wavelet domain. *IEEE Transactions on Image Processing, 12*(11), 1338–1351.
20. Luisier, F., Blu, T., & Unser, M. (2007). A new SURE approach to image denoising: Interscale orthonormal wavelet thresholding. *IEEE Transactions on Image Processing, 16*(3), 593–606.
21. Zhang, L., Dong, W., Zhang, D., & Shi, G. (2010). Two-stage image denoising by principal component analysis with local pixel grouping. *Pattern Recognition, 43*(4), 1531–1549.
22. Elad, M., & Aharon, M. (2006). Image denoising via sparse and redundant representations over learned dictionaries. *IEEE Transactions on Image Processing, 15*(12), 3736–3745.
23. Dong, W., Li, X., Zhang, L., & Shi, G. (2011). Sparsity-based image denoising via dictionary learning and structural clustering. In *Proceedings of IEEE International Conference on Computer Vision Pattern Recognition* (pp. 457–464). Colorado, USA.
24. Mairal, J., Bach, F., Ponce, J., Sapiro, G., & Zisserman, A. (2009). Non-local sparse models for image restoration. In *Proceedings of IEEE International Conference on Computer Vision* (pp. 2272–2279).
25. Goodfellow, I., Pouget-Abadie, J., Mirza, M., Xu, B., Warde-Farley, D., Ozair, S., et al. (2014). Generative adversarial nets. In *Proceedings of Advance in Neural Information Processing System* (pp. 2672–2680). Available: http://papers.nips.cc/paper/5423-generative-adversarial-nets.pdf.

26. Pan, Z., Yu, W., Yi, X., Khan, A., Yuan, F., & Zheng, Y. (2019). Deep multimodal representation learning: A survey. *IEEE Access Digital Object Identifier, 7.* http://doi.org/10.1109/ACCESS.2019.2905015.

27. LeCun, Y., Bottou, L., Bengio, Y., & Haffner, P. (1998). Gradient-based learning applied to document recognition. *Proceedings of the IEEE, 86*(11), 2278–2324.

28. Pan, Z., Yu, W., Yi, X., Khan, A., Yuan, F., & Zheng, Y. (2019) Recent progress on generative adversarial networks (GANs): A survey. *IEEE Access, 7,* 36322–36333. https://doi.org/10.1109/access.2019.2905015.

29. Radford, A., Metz, L., & Chintala, S. (2016). Unsupervised representation learning with deep convolutional generative adversarial networks. In *Proceedings of International Conference on Learn Representation* (p. 116). Available: https://arxiv.org/abs/1511.06434.

30. Mirza, M., & Osindero, S. (2014). Conditional generative adversarial nets. Available: https://arxiv.org/abs/1411.1784.

31. Makhzani, A., Shlens, J., Jaitly, N., & Goodfellow, I. (2016). Adversarial autoencoders. In *Proceedings of International Conference on Learning Representation* (pp. 1–16). Available: http://arxiv.org/abs/1511.05644.

32. Donahue, J., Krähenbühl, P., & Darrell, T. (2017). Adversarial feature learning. In *Proceedings of International Conference Learning Representation* (pp. 1–18). Available: https://arxiv.org/abs/1605.09782.

33. Dumoulinet, V., et al. (2017). Adversarially learned inference. In *Proceedings of International Conference on Learning Representation* (pp. 1–18). Available: https://arxiv.org/abs/1606.00704.

34. Ulyanov, D., Vedaldi, A., & Lempitsky, V. (2018). It takes (only) two: Adversarial generator-encoder networks. In *Proceedings of AAAI Conference on Artificial Intelligence.* Available: https://www.aaai.org/ocs/index.php/AAAI/AAAI18/paper/view/16568.

35. Metz, L., Poole, B., Pfau, D., & Sohl-Dickstein, J. (2017). Unrolled generative adversarial networks. In *Proceedings of International Conference on Learning Representation* (pp. 1–25). Available: https://arxiv.org/abs/1611.02163.

36. Arjovsky, M., Chintala, S., & Bottou, L. (2017). Wasserstein generative adversarial networks. In *Proceedings of International Conference on Machine Learning* (Vol. 70, pp. 214–223). Available: http://proceedings.mlr.press/v70/arjovsky17a.html.

37. Gulrajani, I., Ahmed, F., Arjovsky, M., Dumoulin, V., & Courville, A. C. (2017). Improved training of Wasserstein GANs. In *Proceedings of 30th Advances in Neural Information Processing System* (pp. 5767–5777). Available: http://papers.nips.cc/paper/7159-improved-training-of-wassersteingans.pdf.

38. Petzka, H., Fischer, A., & Lukovnicov, D. (2018). On the regularization of Wasserstein GANs. In *Proceedings of International Conference on Learning Representation* (p. 124). Available: https://arxiv.org/abs/1709.08894.

39. Zhao, J., Mathieu, M., & LeCun, Y. (2017). Energy-based generative adversarial network. In *Proceedings of International Conference on Learning Representation* (pp. 1–17). Available: https://arxiv.org/abs/1609.03126.

40. Shin, Y. G., Sagong, M. C., Yeo, Y. J., & Ko, S. J. (2019). Adversarial context aggregation network for low-light image enhancement. *Digital Image Computing: Techniques and Applications (DICTA), IEEE.* http://doi.org/10.1109/DICTA.2018.8615848. January 17 2019.

41. Swami, K., & Das, S. K. (2018). Candy: Conditional adversarial networks based fully end-to-end system for single image haze removal. *Computer Vision and Pattern Recognition 2018.* arXiv:1801.02892.

42. Zhang, H., & Patel, V. M. (2018). Densely connected pyramid dehazing network. *Computer Vision and Pattern Recognition (cs.CV) 2018.* arXiv:1803.08396.

43. Sim, H., Ki, S., Choi, J. S., Kim, S. Y., Seo, S., Kim, S., & Kim, M. (2018). High-resolution image dehazing with respect to training losses and receptive field sizes. *IEEE Computer Vision and Pattern Recognition Workshops (CVPRW),* 1025–1032, June 18, 2018.

44. Nguyen, V. G., & Nguyen, D. L. (2019). Joint image deblurring and binarization for license plate images using deep generative adversarial networks. In *5th NAFOSTED Conference on*

Information and Computer Science (NICS), IEEE. https://doi.org/10.1109/nics.2018.8606802, January 10, 2019.

45. Zhang, H., Sindagi, V., & Patel, V. M. (2017). Image de-raining using a conditional generative adversarial network. Available: https://arxiv.org/abs/1701.05957.

46. Li, Z., Zhang, J., Fang, Z., & Huang, B. (2019). Single image snow removal via composition generative adversarial networks. *IEEE Access, 99,* 1. https://doi.org/10.1109/access.2019. 2900323.

47. Li, T., & Lun, D. P. (2019). Single-image reflection removal via a two-stage background recovery process. *IEEE Signal Processing Letters, 26*(8).

48. Yang, G., Yu, S., Dong, H., Slabaugh, G., Dragotti, P. L., Ye, X., et al. (2018). DAGAN: Deep de-aliasing generative adversarial networks for fast compressed sensing mri reconstruction. *IEEE Transactions on Medical Imaging, 37*(6). https://doi.org/10.1109/tmi.2017. 2785879, December 21, 2017.

49. Yu, A., Liu, S., Wei, X., Fu, T., & Liu, D. (2019). Generative adversarial networks with dense connection for optical coherence tomography images denoising. In *11th International Congress on Image and Signal Processing, BioMedical Engineering and Informatics (CISP-BMEI).* https://doi.org/10.1109/cisp-bmei.2018.8633086, February 04, 2019.

50. Yang, Q., Yan, P., Zhang, Y., Yu, H., Shi, Y., Mou, X., et al. (2018). Low-dose ct image denoising using a generative adversarial network with wasserstein distance and perceptual loss. *IEEE Transactions on Medical Imaging, 37*(6). https://doi.org/10.1109/tmi.2018.2827462, April 17, 2018.

51. Lyu, Y., Jiang, W., Lin, Y., Voros, L., Zhang, M., Mueller, B., et al. (2019). Motion-blind blur removal for CT images with wasserstein generative adversarial networks. In *11th International Congress on Image and Signal Processing, BioMedical Engineering and Informatics (CISP-BMEI), IEEE.* https://doi.org/10.1109/cisp-bmei.2018.8633203, February 04, 2019.

52. Mishra, D., Chaudhury, S., Sarkar, M., Soin, A. S. (2018). Ultrasound image enhancement using structure oriented adversarial network. *IEEE Signal Processing Letters, 25*(9).

53. Park, H. S., Baek, J., You, S. K., Choi, J. K., & Seo, J. K. (2019). Unpaired image denoising using a generative adversarial network in X-Ray CT. *IEEE Access, 7.* https://doi.org/10.1109/ access.2019.2934178, August 09, 2019.

54. Du, W., Chen, H., Liao, P., Yang, H., Wang, G., & Zhang, Y. (2019). Visual attention network for low dose ct. https://doi.org/10.1109/lsp.2019.2922851, arXiv:1810.13059.

55. Huang, D., Hou, C., Yang, Y., Lang, Y., & Wang, Q. (2018). Micro-doppler spectrogram denoising based on generative adversarial network. In *48th European Microwave Conference (EuMC), IEEE.* https://doi.org/10.23919/eumc.2018.8541507, November 22, 2018.

56. Wang, P., Zhang, H., & Patel, V. M. (2018). Generative adversarial network-based restoration of speckled SAR images. In *IEEE 7th International Workshop on Computational Advances in Multi-Sensor Adaptive Processing (CAMSAP).* https://doi.org/10.1109/camsap.2017.8313133, March 12, 2018.

57. Chen, Y., Wu, F., & Zhao, J. (2018). Motion deblurring via using generative adversarial networks for space-based imaging. In *IEEE 16th International Conference on Software Engineering Research, Management and Applications (SERA).* https://doi.org/10.1109/sera.2018.8477191.

58. Liu, J., Sun, W., & Li, M. (2018). Recurrent conditional generative adversarial network for image deblurring. *IEEE Access, 7.* https://doi.org/10.1109/access.2018.2888885, December 21, 2018.

59. Wu, S., Fan, T., Dong, C., & Qiao, Y. (2019). RDS-denoiser: a detail-preserving convolutional neural network for image denoising. *IEEE International Conference on Cyborg and Bionic Systems (CBS).* https://doi.org/10.1109/cbs.2018.8612215, January 17, 2019.

60. Nuthna, V., Chachadi, K., & Joshi, L. S. (2019). Modeling and performance evaluation of generative adversarial network for image denoising. In *International Conference on Computational Techniques, Electronics and Mechanical Systems (CTEMS), IEEE.* https://doi.org/10. 1109/ctems.2018.8769231, July 25, 2019.

61. Chen, L., Dan, W., Cao, L., Wang, C., & Li, J. (2018). Joint denoising and super-resolution via generative adversarial training. In *24th International Conference on Pattern Recognition (ICPR).* https://doi.org/10.1109/icpr.2018.8546286, August 2018.

62. Alsaiari, A., Rustagi, R., Thomas, M. M., & Forbes, A. G. (2019). Image denoising using a generative adversarial network. In *IEEE 2nd International Conference on Information and Computer Technologies (ICICT), IEEE*. https://doi.org/10.1109/infoct.2019.8710893, May 13, 2019.

63. Chen, J., Chen, J., Chao, H., & Yang, M. (2018). Image blind denoising with generative adversarial network based noise modeling. In *IEEE/CVF Conference on Computer Vision and Pattern Recognition*. https://doi.org/10.1109/cvpr.2018.00333, December 17, 2018.

64. Creswell, A., & Bharath, A. A. (2019). Denoising adversarial autoencoders. *IEEE Transactions on Neural Networks and Learning Systems, 30*(4).

65. ZhiPing, Q., YuanQi, Z., Yi, S., & XiangBo, L. (2018). A new generative adversarial network for texture preserving image denoising. In *Eighth International Conference on Image Processing Theory, Tools and Applications (IPTA), IEEE*. https://doi.org/10.1109/ipta.2018.8608126, January 14, 2019.

66. Yu, F., & Koltun, V. (2016). Multi-scale context aggregation by dilated convolutions. In *Proceedings of the International Conference on Learning Representations*.

67. AAPM. (2017). Low dose CT grand challenge. Available: http://www.aapm.org/GrandChallenge/LowDoseCT/#.

68. Hennig, J., Nauerth, A., & Friedburg, H. (1986). RARE imaging: A fastimaging method for clinical MR. *Magnetic Resonance in Medicine, 3*(6), 823–833.

69. Carnegie Mellon university motion capture database. http://mocap.cs.cmu.edu/. Accessed February 13, 2019.

70. Russakovsky, O., Deng, J., Su, H., et al. (2015). Imagenet large scale visual recognition challenge. *International Journal of Computer Vision, 115*(4), 211–252.

71. Roth, S., & Black, M. J. (2009). Fields of experts. *International Journal of Computer Vision, 82*(2), 205.

72. Paltz, T., & Roth, S. (2017). Benchmarking denoising algorithms with real photographs. In *2017 IEEE Conference on Computer Vision and Pattern Recognition, CVPR 2017* (pp. 2750–2759). Honolulu, HI, USA, July 21–26, 2017.

73. Nah, S., Kim, T. H., & Lee, K. M. (2017). Deep multi-scale convolutional neural network for dynamic scene deblurring. In *Proceedings of IEEE Conference on Computer Visual and Pattern Recognition* (pp. 257–265).

74. Timofte, R., Agustsson, E., Van Gool, E., et al. (2017). Nature 2017 challenge on single image super-resolution: Methods and results. In *2017 IEEE Conference on Computer Vision and Pattern Recognition Workshops (CVPRW)* (pp. 1110–1121).

75. Zhang, L., Wu, X., Buades, A., et al. (2011). Color demosaicing by local directional interpolation and nonlocal adaptive thresholding. *Journal of Electronic Imaging, 20*(2), 023016.

76. Zeyde, R., Elad, M., Protter, M. (2010). On single image scale-up using sparse representations. In *International conference on curves and surfaces* (pp. 711–730). Springer, Berlin, Heidelberg.

77. Lebrun, M., Colom, M., & Morel, J. M. (2015). The noise clinic: A blind image denoising algorithm. *Image Processing On Line, 5,* 1–54.

Deep Learning-Based Lossless Audio Encoder (DLLAE)

Uttam Kr. Mondal, Asish Debnath, and J. K. Mandal

Abstract Lossless audio compression is a crucial technique for reducing size of audio file with preservation of audio data. In this current approach, a lossless audio encoder has been designed with the help of deep learning technique, followed by entropy encoding to generate compressed encoded data. Nine hidden layers have been considered in the proposed network for the present encoding framework. Experimental results are shown with statistical parameters for comparing the performance and quality of the current approach with other standard algorithms.

Keywords Neural network · Lossless audio compression · Audio sampling · Convolution · Deep learning · Entropy encoder

1 Introduction

Lossless compression [1] is a technique for compression which is used when the requirement is to keep the original and the decompressed data are identical. Moreover, it reduces the file size without any deviation of content from the original data. Neural network [2] represents data in concise and effective way for processing and its architecture helps to represent sampled audio in more concise and cohesive way. Deep learning [3] technology is successfully applied in various computer vision applications [4] for similar structured representations of data structure. Therefore, it is chosen to design audio encoder as a framework.

U. Kr. Mondal (✉)
Department of Computer Science, Vidyasagar University, Midnapore, West Bengal, India
e-mail: uttam_ku_82@yahoo.co.in

A. Debnath
Tata Consultancy Services Limited, Kolkata, West Bengal, India
e-mail: debnathasish@gmail.com

J. K. Mandal
Department of Computer Science and Engineering, University of Kalyani, Nadia, West Bengal, India
e-mail: jkm.cse@gmail.com

© Springer Nature Singapore Pte Ltd. 2020
J. K. Mandal and S. Banerjee (eds.), *Intelligent Computing: Image Processing Based Applications*, Advances in Intelligent Systems and Computing 1157,
https://doi.org/10.1007/978-981-15-4288-6_6

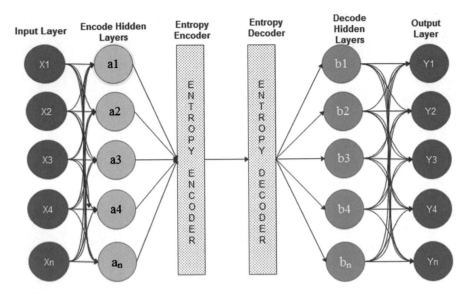

Fig. 1 Architecture of the deep learning-based system

In this paper, the framework is designed in three steps. First, the audio signal is converted to digital signal from analog signal based on existing Nyquist criteria. In second step, audio sampled values are divided into separate groups. Each group consists of five elements. Sampled values in each group are processed sequentially. In each layer, group of five elements is reduced to four-transformed elements. So, one element reduced during the compression time from each group of layers. This process applied iteratively on every group of a layer. After processing all the groups, encoded form of data passed to the next encoding hidden layer. In the last step, encoded data stream from 9th hidden layer is passed to entropy encoder. Finally, encoded stream is produced from the output of the entropy encoder. The architecture of the deep learning [5] based encoder framework is shown in Fig. 1.

The organization of the paper as follow: Sect. 2 describes the overall technique of the proposed methodology. The encoding algorithm and decoding algorithms are discussed in Sects. 2.1 and 2.2, respectively. The experimental results are depicted in Sect. 3. Conclusion is drawn in Sect. 4 whereas references at the end.

2 The Technique

The overall technique defines the audio signal is compressed with the help of encoding in the hidden layers of the deep learning network followed by entropy encoder shown in Fig. 1. The details of the technique are depicted in Fig. 2.

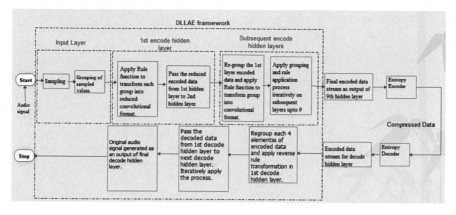

Fig. 2 Block diagram of encoding process and decoding process

DLLAE–ESV algorithm contains the compression technique, elaborated in Sect. 2.1. DLLAE–ESV describes how the audio sampled values are compressed in hidden layers. Further, entropy encoder applied on this bit stream. DLLAE–DCD, the decompression algorithm discussed in the Sect. 2.2. Decoding of encoded technique is shown in Sect. 2.2.

The groupwise encoding technique explained below with the help of an example: Let a group, say Gr1, as below:

$$Gr1 = \begin{bmatrix} 4 \\ 3 \\ 2 \\ 7 \\ 9 \end{bmatrix}$$

Let $m = 4$, $n = 3$, $o = 2$, $p = 7$, $q = 9$
Consider Gr1′ as the encoded group.
The average of Gr1, i.e., $r = 5$.
Divide the first element of the group Gr1(i.e., 4) into three equal parts. The value of each part is

$$x = \frac{4}{3} = 1.3333$$

Add this x (1.3333) with 3rd, 4th, and 5th element.
Therefore, the transformed 3rd, 4th, and 5th elements are 3.3333, 8.3333, and 10.3333, respectively.
Remove the 1st element of the Gr1 (4) from the Gr1′. So, 1st element of Gr1′ will be the average value (i.e., 5). 2nd, 3rd, and 4th elements of Gr1′ will be 3.3333, 8.3333, and 10.3333, respectively.

Finally, the Gr1′'s will be as below:

$$Gr1' = \begin{bmatrix} 5 \\ 3.3333 \\ 8.3333 \\ 10.3333 \end{bmatrix}$$

Let $o' = 3.3333, p' = 8.3333, q' = 10.3333$. o', p', and q' are transformed values. Decoding process is explained as below:

$$m = 3 * x = 4$$

$$o' = o + x$$

$$o = o' - x = 3.3333 - 1.3333 = 2$$

$$p' = p + x$$

$$p = p' - x = 8.3333 - 1.3333 = 7$$

$$q' = q + x$$

$$q = q' - x = 10.3333 - 1.3333 = 9$$

$$r = \frac{m + n + o + p + q}{5}$$

$$n = 5r - m - o - p - q = 25 - 4 - 2 - 7 - 9 = 3$$

i.e., the decoded group $= \begin{bmatrix} 4 \\ 3 \\ 2 \\ 7 \\ 9 \end{bmatrix}$

Therefore, the encoded and decoded group elements are in sync.

2.1 Encoding of Sampled Values (DLLAE–ESV)

The **DLLAE–ESV** compression algorithm starts with audio signal. Encode and compress the sampled values using neural network-based architecture. Encoded data stream further compressed using entropy encoder. Algorithm 1 describes the encoding technique.

Algorithm 1 DLLAE–ESV

Input: A slice of audio signal.
Output: Encoded audio signal.
Method: The encoding procedure is described in following steps:
Step 1 (Sampling): To convert the analog audio signal (input) to digital signal, sampling process is applied to the audio signal. As per the Nyquist criteria, let, f_s is the sampling frequency and f_{max} be the highest frequency component in the audio signal's magnitude spectrum. f_s and f_{max} related as below equation:

$$f_s > 2f_{max}$$

Step 2 (Grouping): Segregate the sampled values into separate groups. Each group will contain five entities:

$$\text{Total number of groups in first hidden layer} = \frac{\text{Total number of original sampled values}}{5}$$

Step 3 (Encoding in deep neural net): Apply transformational following rules to transform the sampled values into encoded form.

i.   Select the first group (Gr1) from the input layer.

$$\text{Let, Gr1} = \begin{bmatrix} m \\ n \\ o \\ p \\ q \end{bmatrix}$$

ii. Find average of (Gr1), i.e., divide the five elements with 5.

Let, f be denoted by average of 5 elements of a group 1

$$r = \frac{m + n + o + p + q}{5}$$

iii. Divide the first-sampled value of Gr1 into three equal parts.

Let each equal part is denoted by x.

$$x = \frac{m}{3}$$

iv. Distribute these three parts with the last 3 entities of group 1. Means, add each part with 3rd, 4th, and 5th sentities of Gr1

$$o' = o + x$$

$$p' = p + x$$

$$q' = q + x$$

Here o', p', and q' are the encoded values of c, d, and e.

v. Remove the first-sampled values from the Gr1 vector and replace the second value with the average value, i.e., f.

vi. Encoded representation of Gr1 in the first-hidden layer will be as below:

$$\text{Gr1}' = \begin{bmatrix} r \\ o' \\ p' \\ q' \end{bmatrix}$$

vii. Same steps [i–vi] are iteratively applied on all the groups of input layer and represented in the reduced format in the first-hidden layer.

Step 4: Steps 2–3 iteratively applied on successive encode hidden layers. Intermediate compressed form of data is available as the output of the 9th encode hidden layer.

Step 5: Entropy encoder is applied to the encoded data stream generated at the end of step 4 of Algorithm 1 for further compression.

2.2 Decoding of Encoded Data (DLLAE–DCD)

Algorithm 2 describes the decompression technique. The decompression algorithm reconstructs the original audio from the encoded data stream.

Algorithm 2 DLLAE–DCD

Input: Compressed and encoded stream.
Output: Reconstructed audio signal.
Method: The decompression is described as below steps:
Step 1 (Entropy decoder): Entropy decoding technique is applied on the encoded stream.
Step 2 (Decompression): Output of the entropy decoder will be taken as an input to the innermost hidden layer of decompression, i.e., hidden decode layer 1. Each group of the output of the 4th hidden layer contains four entities.

Let, consider the first transformed group is Gr1'.

$$Gr1' = \begin{bmatrix} r \\ o' \\ p' \\ q' \end{bmatrix}$$

Below are the five equations:

$$m + n + o + p + q = 5r \tag{1}$$

$$m = 3x \tag{2}$$

$$o' = o + x \tag{3}$$

$$p' = p + x \tag{4}$$

$$q' = q + x \tag{5}$$

From Eqs. (1)–(5), values of $m, n, o, p,$ and q are to be decoded.
From Eqs. (1) and (2), we get

$$3x + n + o + p + q = 5r \tag{6}$$

From Eq. (3), we calculate the value of o as the value of o' is known.

$$o = o' - x$$

From Eq. (4), we calculate the value of p as the value of p' is known.

$$p = p' - x$$

From Eq. (5), we calculate the value of q as the value of q' is known.

$$q = q' - x$$

Now, substitute the value of $o, p, q,$ and r in Eq. (6). r is known value.

$$3x + n + o' - x + p' - x + q' - x = 5r$$

$$n = 5r - o' - p' - q'$$

Finally, original value of n is calculated. Therefore, $m, n, o, p,$ and q values are known.

Now, the original Gr1 will be as below:

$$Gr1 = \begin{bmatrix} m \\ n \\ o \\ p \\ q \end{bmatrix}$$

Step 3: Step 1 iteratively applied to all the groups of innermost hidden decoding layer 1 and represented in the decoded format.

Step 4: Step 2 iteratively applied on second, third, and final decoding hidden layer. Final decompressed form of data is available as the output of the fourth and final decoding hidden layer.

3 Results and Analysis

Data compression ratio is the ratio between the uncompressed size and compressed size file,

$$\text{Compression Ratio} = \frac{\text{Uncompressed Size}}{\text{Compressed Size}} \tag{7}$$

Space savings is another crucial compression performance measurement factor, which is represented in Eq. (8):

$$\text{Space savings} = 1 - \frac{\text{compressed size}}{\text{uncompressedsize}} \tag{8}$$

The details of the ten songs used for experiment are present in Table 1. The compression ratio for the selected songs compare with other standard algorithms Monkey's Audio [6], WavPack Lossless [7], FLAC lossless using Fonepaw [8] is present in Table 2 whereas the details of considering song signals are shown in Table 1. Figure 3 shows the difference between the original audio and the regenerated audio signals. Observing the compression ratio of Table 2, DLLAE represents the highest compression ratio than existing standard techniques. It also provides the highest compression ratio for each individual group (i.e., Pop, Sufi, Rock, Rabindra Sangeet, and Classical).

4 Conclusion

In this paper, a neural network-based lossless audio encoder is proposed and implemented. Present concept achieves more than 90% compression without losing any audio information. The performance of the algorithm may also be improved by

Table 1 Details of the audio songs used for the experiment like number of channels, sample rate, total samples, and bits per sample

Audio	Number of channels	Sample rate	Total samples	Bits per sample
Rabi_1.wav	1	44,100	433,368	16
Rabi_2.wav	1	48,000	514,832	16
Classical_1.wav	1	48,000	489,712	16
Classical_2.wav	1	48,000	524,832	16
Pop_1.wav	1	44,100	477,968	16
Pop_2.wav	1	48,000	484,592	16
Sufi_1.wav	1	44,100	495,440	16
Sufi_2.wav	1	44,100	455,440	16
Rock_1.wav	1	48,000	470,496	16
Rock_2.wav	1	48,000	460,256	16

searching better encoding rules. Performance improvement of the proposed system is further scope of the work by using more training. Groupwise data encoding inside neural network could be processed parallel for further performance enhancement. Automatic layer selection based on the required compression ratio is future scope of the work.

Table 2 Compression ratio (%) for five different types of songs for Monkey's audio [6], WavPack lossless [7], FLAC lossless using Fonepaw [8], and DLLAE

Song type	Audio files (.wav) [10 s]	Compression ratio / Techniques											
		Monkey's audio [6]			WavPack lossless [7]			FLAC using Fonepaw [8]			DLLAE		
		Individual (%)	Avg. (%)	Over all avg. (%)	Individual (%)	Avg. (%)	Over all avg. (%)	Individual (%)	Avg. (%)	Over all avg. (%)	Individual (%)	Avg. (%)	Over all avg. (%)
Rabindra Tagore	Rabi_1	53.48	53.95	57.65	40.66	40.62	54.98	72.77	72.75	61.41	91.39	91.35	91.7
	Rabi_2	54.39			40.58			72.73			91.40		
Classical	Classical_1	62.99	63.55		28.91	28.44		68.69	68.19		92.45	91.94	
	Classical_2	64.12			27.97			67.70			91.44		
Rock	Rock_1	61.11	62.14		32.92	31.24		69.56	69.19		92.44	91.45	
	Rock_2	63.17			29.57			68.82			90.47		
Pop	Pop_1	53.93	57.32		41.37	35.99		72.74	71.45		93.37	92.41	
	Pop_2	60.71			30.61			70.17			91.45		
Sufi	Sufi_1	49.66	51.29		47.30	45.23		75.13	74.32		92.44	91.35	
	Sufi_2	52.93			43.16			73.52			91.43		

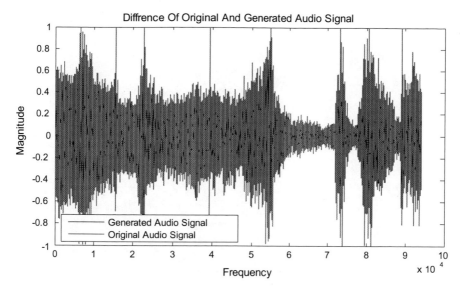

Fig. 3 Deviations of original and regenerated audio signals

References

1. Mondal, U. K. (2019). Achieving lossless compression of audio by encoding its constituted components (LCAEC). *Innovations in Systems and Software Engineering, 15,* 75–85. https://doi.org/10.1007/s11334-018-0321-x.
2. Parisi, G., Kemker, R., Part, J., Kanan, C., & Wermter, S. (2018). Continual lifelong learning with neural networks: A review. *Neural Networks.* https://doi.org/10.1016/j.neunet.2019.01.012.
3. Cheng, Z., et al. (2018). Deep convolutional autoencoder-based lossy image compression, picture coding symposium. In *PCS 2018—Proceedings,* ISBN (Print)-9781538641606, 253–257.
4. Buda, M., Maki, A., & Mazurowski, M. A. (2017) A systematic study of the class imbalance problem in convolutional neural networks. *Neural Networks, 106.* https://doi.org/10.1016/j.neunet.2018.07.011.
5. Srivastava, N., et al (2014) Dropout: A simple way to prevent neural networks from overfitting. *Journal of Machine Learning Research, 15,* 1929–1958.
6. https://www.monkeysaudio.com/ accessed on 15.09.2019 at 11 AM.
7. https://www.wavpack.com/ accessed on 14.08.2019 at 9 AM.
8. https://fonepaw-video-converter-ultimate.en.softonic.com/. Accessed January 16, 2019 at 10 PM.

A New Repeated Pixel Value Difference-Based Steganographic Scheme with Overlapped Pixel

Partha Chowdhuri, Pabitra Pal, Biswapati Jana, and Debasis Giri

Abstract The steganographic scheme based on pixel value difference (PVD) embeds secret information by modifying the value of pixel pair within the block. The embedding capacity, visual quality, and performance of the PVD-based algorithm depend on the correlation between the pixel pair. To improve the capacity, quality, and performance of PVD-based algorithm, a new steganographic scheme has been proposed which uses overlapped pixels of a block to embed secret data within the R,G,B color components of interpolated pixel of a color image. At first, the RGB color components of the host image are partitioned into non-overlapped pixel blocks. After that, image interpolation is applied to enlarge each block. Then the secret message is embedded within the interpolated pixels using PVD algorithm. It has been observed that the new approach provides 2.22 bpp payload with 43.47 (dB) PSNR. The proposed scheme has been tested through different steganographic attacks. The experimental results show that the proposed algorithm has better embedding capacity compared to the conventional PVD-based approach.

Keywords Steganography · PVD · Payload · PSNR · Image interpolation

P. Chowdhuri · P. Pal · B. Jana (✉)
Department of Computer Science, Vidyasagar University, West Midnapore 721102, India
e-mail: biswapatijana@gmail.com

P. Chowdhuri
e-mail: prc.email@gmail.com

P. Pal
e-mail: pabipaltra@gmail.com

D. Giri
Department of Information Technology, Maulana Abul Kalam Azad University of Technology,
Haringhata, Nadia, West Bengal 741249, India
e-mail: debasis_giri@hotmail.com

© Springer Nature Singapore Pte Ltd. 2020
J. K. Mandal and S. Banerjee (eds.), *Intelligent Computing: Image Processing
Based Applications*, Advances in Intelligent Systems and Computing 1157,
https://doi.org/10.1007/978-981-15-4288-6_7

1   Introduction

Steganography is the art of hidden communication using a multimedia document like image, audio, or video. To embed the secret data, we have to change the pixel values of the cover image. The aim of such embedding is that, the resulting stego image will be visually almost similar to the original image. Due to the visual similarity, it will be difficult to detect the presence of a secret data in the stego image. Till date, many steganographic schemes [1–11] have been developed including Wu and Tsai's [12] PVD in 2003, that uses the difference of two adjacent pixels to hide secret data. The bits that can be embedded depends on the difference between the pixels pair. The higher difference can embed more bits and vice versa. The difference between the pixels pair will be higher in the edge area of an image. So, more data bits can be embedded in the edge area compared to smooth area of any image. A range table is used in PVD method to calculate the number of embedding bits within the pixel pair. Each sub-range in the range table is a power of 2. The difference value of the pixel pair is mapped onto the range table. The number of bit to be embedded is calculated using some specific formula depending on the range table. The main drawback of this scheme is that the overflow and underflow problem can arise here. Most of the data hiding methods developed so far using PVD are not reversible. So, we have proposed a new PVD technique to hide data in a color image that not only preserves the visual quality of the image and gives high embedding capacity but also fully reversible. Also, the Chi-square analysis, RS analysis, and normalized cross-correlation (NCC) show better results than other existing methods.

The rest of the paper is constructed as follows: In Sect. 2, various related works have been listed. Motivation and objective of our proposed technique are discussed in Sect. 3. In Sect. 4, the proposed method is discussed. Experimental results and comparisons are shown in Sect. 5. Finally, conclusions are drawn in Sect. 6.

2   Related Work

The least significant bit (LSB) substitution method is one of the most popular and earliest methods of image steganography. It does not change the size of an image and replaces only the LSB according to the algorithmic needs. Though the computational complexity of the LSB method is very low, the visual quality of the image is degraded when more LSBs are modified. The embedding method is also revealed by various developments of steganalysis. One of the popular approaches for data hiding is done by pixel value difference (PVD) which is based on the difference between adjacent pixel pair. In PVD, the embedding capacity can be increased if the difference between pixel pair is increased. Our vision systems are less sensitive to changes in higher texture or edge areas of an image but it is more sensitive to changes in smooth areas of an image. The PVD method holds this property to hide less data in smooth areas and more data in edge areas of images. Wu and Tsai [12] first proposed a pixel value difference (PVD) method. Since then, various PVD-based data hiding methods have been developed to enhance security, increase data hiding capacity, and improve visual quality of the stego image. Wang et al. [13] advised another PVD method that not only removes the overflow and underflow problem but also ensures better visual quality of the stego image. For minimum distortion, they compute remainder of two pixel pairs using a modulus function instead of simple pixel value difference. Although the embedding capacity and the visual quality of the stego image are in the acceptable level of current market demand, the existence of the secret message can easily be detected by histogram analysis. Joo et al. [14] designed a PVD method which uses a turnover policy to prevent abnormal increases and fluctuations in the histogram values to ensure better security. Liao et al. [15] formulated a technique which utilizes PVD for classifying the smooth and edge areas of four-pixel blocks to get high visual quality and high payload. Then, Lee et al. [16] developed an embedding method with high payload by tri-way PVD using JPEG2000 image compression on the secret image. Liao et al. [17] suggested a high payload embedding method which is based on four-pixel differencing to get good visual quality of the image. Khodaei et al. [18] proposed a PVD technique which targeted improved payload but their method gives poor visual quality and suffers from histogram analysis attack. Tseng et al. [5] suggested a steganographic scheme using PVD and the perfect square number. Though in this scheme they achieved a very good PSNR value, they have to compromise with payload. Balasubramanian et al. [19] developed an octonary PVD scheme with (3 × 3) pixel blocks to enhance security and achieve high payload. In the same year, Chen et al. [20] proposed a PVD-based scheme to obtain better visual quality with improved security. After that, Jung and Yoo [21] designed a data hiding method which is based on index function. This method compromises the image quality to improve the data hiding capacity and maintain the histogram as it is. Shen and Huang [22] suggested a PVD method exploiting modification directions (EMD). This method obtains high payload, and it is undetectable under PVD histogram analysis. In the next year, Swain [23] used adaptive PVD method using horizontal and vertical edges for color images. They proposed two separate methods with non-overlapped and overlapped image pixel blocks. Compared to Wu and Tsai [12] method, this method has low payload but it has good visual quality. It resists RS analysis attack and difference histogram analysis attack. After that, Swain [8] proposed adaptive and non-adaptive PVD steganography using overlapped pixel blocks. This scheme provides higher embedding capacity as compared to the existing non-adaptive PVD techniques. To increase the embedding capacity, Swain [11] proposed a scheme. In this scheme, Swain uses quotient value differencing and LSB substitution technique. Lutovac et al. [24] developed an algorithm for robust watermarking in DCT domain using Zernike image moments. Hussain et al. [25] evolved a data hiding scheme that uses parity-bit pixel value differencing to improve embedding capacity. In the year 2018, Jung [26] germinated another data hiding scheme that improves payload using mixed PVD and LSB on bit plane. Dorfer et al. [27] suggested a paper on end-to-end cross-modality retrieval with CCA projections and pairwise ranking loss. Nagai et al. [28] provided a nobel digital watermarking techniques for deep neural networks. Hameed et al. [6] proposed a new data hiding technology using adaptive directional pixel value differencing method. They achieve a high embedding capacity with better PSNR. Sahu and Swain [9] proposed a new hiding approach using PVD and modulus function. Luo et al. [10] suggested new steganographic technique based on PVD named, traceable quantum steganography. Their scheme mainly follows image edge effects and human visual system characteristics very well.

Generally, a PVD-based data hiding method provides high visual quality due to varying pixel value difference and possible to improve capacity while keeping it undetectable under various attacks. In this literature, several researchers apply PVD-based steganographic scheme but they ignored color pallet separation. Again, PVD was not reversible.

3 Motivation and Objective

PVD-based data hiding schemes proposed so far can only hide limited secret bits within a single block. Moreover, PVD-based schemes are not reversible. So, our main objective is to design a new PVD-based reversible scheme to increase the embedding capacity while keeping stable good visual quality than other existing schemes. Securing the scheme from several steganographic attacks is also another important objective of our proposed scheme. The contributions of the paper are mentioned below:

- Till date, only one PVD operation has been performed using a pair of pixel within a block. We have developed a new approach, where a single pixel is participated multiple times in repeated PVD operation.
- The data extraction through one PVD operation was investigated by the previous researchers, but it is hard to retrieve the hidden information from repeated PVD operation with overlapped pixel. Here we have suggested an innovative scheme to successfully retrieve the secret information from repeated overlapped pixel blocks.
- In this scheme, three color components are used to hide the secret message. It helps to minimize the distortion in the stego image by hiding secret information in separate color pallet which are less sensitive of human visual system.
- In most of the cases, RED block remains unchanged due to the conversion of d' in 8-bit form. But, major change is made in BLUE block which is less sensitive in our human eyes.

4 Proposed Steganographic Scheme

4.1 Data Embedding Procedure

The embedding process of the proposed scheme is shown in Fig. 1. PVD method is not reversible. So, during embedding, if the pixels are changed, they cannot get back their original values after secret message extraction. For this reason, to keep

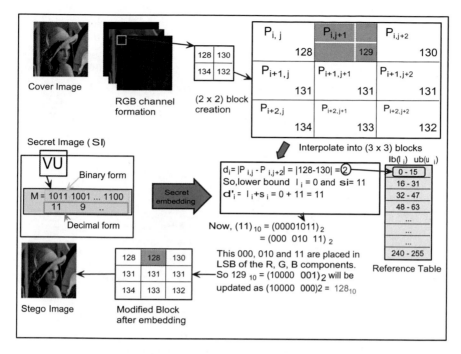

Fig. 1 Numerical illustration of embedding process

the original pixels unchanged, image interpolation is used in this technique to create dummy interpolated pixels where the data can be embedded. At first, the color cover image (C) is split into three separate RED, GREEN and BLUE color components. Then, a (2 × 2) pixel block is selected from the RED color component (shown in Fig. 1). This (2 × 2) pixel block is then converted into (3 × 3) RED color pixel block using image interpolation by Eq. 1. This is repeated for the rest (2 × 2) pixel blocks of the RED color component. This process is iterated for the GREEN and BLUE color components to get the final interpolated color image. On the other hand, the secret message is formed by extracting the color bits from secret color image such as logo image (S) and merging into a bit stream in a sequential order. The secret message is considered as $M = m_1|m_2|m_3|\ldots m_n$, where m_i, $(i = 1$ to $n)$ is 4-bit secret message block. Now take a (3 × 3) RED pixel block from the interpolated image, where $i, j = 0, 1, 2$. Then the pixels of this (3 × 3) block can be considered as $P_{i,j}$, $P_{i,j+1}$, $P_{i,j+2}$, $P_{i+1,j}$, $P_{i+1,j+1}$, $P_{i+1,j+2}$, $P_{i+2,j}$, $P_{i+2,j+1}$, $P_{i+2,j+2}$ as shown in Fig. 1. At first, a pixel pair $P_{i,j}$ and $P_{i,j+2}$ are considered. The absolute difference d_i is calculated as $d_i = |P_{i,j} - P_{i,j+2}|$. Then, 4-bit secret message has been taken and converted into its decimal equivalent sd_i. The subrange of d_i from reference table, as shown in Fig. 1, is checked to get the lower bound l_i of it. Then, l_i is added with sd_i to get d'_i. This d'_i is converted into 8-bit binary number. This 8-bit binary number is separated into three binary bit blocks of length 3, 3, and 2. The first 3 bits are embedded into the LSB + 2, LSB + 1, LSB of the RED color component of pixel $P_{i,j+1}$. The next 3 bits are embedded in the LSB + 2, LSB + 1, LSB of the GREEN color component, and the remaining 2 bits are embedded in the LSB + 1, LSB of the BLUE color component of the pixel $P_{i,j+1}$. Similarly, the pixel pairs $(P_{i,j}, P_{i+2,j})$, $(P_{i,j+2}, P_{i+2,j+2})$, $(P_{i+2,j}, P_{i+2,j+2})$ and $(P_{i,j}, P_{i+2,j+2})$ are considered to embed secret bits into the pixels $P_{i+1,j}$, $P_{i+1,j+2}$, $P_{i+2,j+1}$ and $P_{i+1,j+1}$, respectively. The remaining secret bits are embedded following for the rest of the (3×3) pixel blocks of the interpolated image. Finally, the stego image is created by all updated pixel blocks. The embedding procedure is given in Algorithm 1.

$$CI(i, j) = \begin{cases} C(p, q), \quad \text{where } p = 1, \ldots, M; q = 1, \ldots, N; \\ \dfrac{(C(i, j - 1) + C(i, j + 1))}{2}, \\ \text{where } (i \bmod 2) = 0, (j \bmod 2) \neq 0; i = 1, \ldots, (2 \times M/3); \\ j = 1, \ldots, (2 \times N/3) \\ \dfrac{(C(i - 1, j - 1) + C(i - 1, j + 1) + C(i + 1, j - 1) + C(i + 1, j + 1))}{4}, \\ \text{where } (i \bmod 2) = 0, (j \bmod 2) = 0; i = 1, \ldots, (2 \times M/3); \\ j = 1, \ldots, (2 \times N/3) \end{cases} \quad (1)$$

Algorithm 1: Embedding Algorithm

---

input : Cover image $C_{(M \times N)}$, Secret image S

output: Stego image C'

Algorithm Embedding():
- Step 1: Convert color cover image to an 3 dimensional array, for red, green and blue pixels
- Step 2: Create an interpolated image array
- Step 3: Read the secret image
- Step 4: Extract the secret bits from the secret image array
- Step 5: Embed secret bits in every red, green and blue block of the interpolated image using the **EmbedSecretBits** function

Function EmbedSecretBits(*ImageArray, SecretBits*):
- Step 1: Perform the operation for each 3x3 image blocks
- Step 2: In each iteration embed secret bits in five pixel positions: $P_{i,j+1}$, $P_{i+1,j}$, $P_{i+2,j+1}$, $P_{i+1,j+2}$ and $P_{i+1,j+1}$ as shown in Fig. 1 using **EmbedBitsInPixel** function

Function EmbedBitsInPixel(*ImageArray, SecretBits*):
- Step 1: Get absolute difference value of the pixels
- Step 2: Get the lower bound of the difference
- Step 3: Get decimal value of 4 bit secret data
- Step 4: Add this 4 bit decimal value to the lower bound
- Step 5: Convert this value to 8 bit data.
- Step 6: Convert first 3 bits, second 3 bits and last 2 bits to separate decimal numbers
- Step 7: Change LSB 3 of red pixel, LSB 3 of green and LSB 2 of blue pixel with these values

---

4.2 Data Extraction Procedure

The data extraction procedure is presented in Fig. 2. At first, the stego image is considered as an input image. It is partitioned into (3×3) pixel blocks. Now, the pixels of this (3×3) pixel block can be considered as $P'_{(i,j)}$, $P'_{(i,j+1)}$, $P'_{(i,j+2)}$, $P'_{(i+1,j)}$,

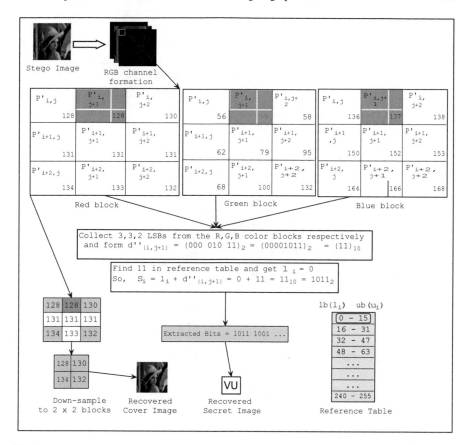

Fig. 2 Numerical illustration of extraction process

$P'_{(i+1,j+1)}$, $P'_{(i+1,j+2)}$, $P'_{(i+2,j)}$, $P'_{(i+2,j+1)}$, and $P'_{(i+2,j+2)}$ as shown in Fig. 2. First, the pixel $P'_{(i,j+1)}$ is considered. The pixel is separated into RGB color components. Then, the 2 LSB of BLUE component, 3 LSB of GREEN component, and 3 LSB of RED component are collected and merged one after another to form a 8-bit binary number. The decimal equivalent of this binary number is stored into $d''_{(i,j+1)}$. The lower bound of the number $d''_{(i,j+1)}$ is taken from the reference table, as shown in Fig. 2 and deducted from $d''_{(i,j)}$ to get m'_1. This m'_1 is actually the secret bits extracted from the pixel $P'_{(i,j+1)}$. Then, in the similar way, the pixels $P'_{(i+1,j)}$, $P'_{(i+1,j+2)}$, $P'_{(i+2,j+1)}$, and $P'_{(i+1,j+1)}$ are considered to get m'_2, m'_3, m'_4, and m'_5 secret message bits. Then m'_1, m'_2, m'_3, m'_4, and m'_5 are merged one after another to get the total secret bits embedded into the first pixel block. The same procedure is followed for the rest of the pixel blocks to get the whole secret message. The secret image is reconstructed from these extracted secret bits. Now, the original (2 × 2) pixel block of the cover image is constructed by eliminating the $P'_{(i,j+1)}$, $P'_{(i+1,j)}$, $P'_{(i+1,j+2)}$, $P'_{(i+2,j+1)}$, and $P'_{(i+1,j+1)}$ pixels from the (3 × 3) pixel block of the input stego image. Then, the same procedure is followed for the rest of the (3 × 3) pixel blocks of the input stego image. Finally, the original cover image is reconstructed from all these recovered (2 × 2) pixel blocks. The extraction procedure is given in Algorithm 2.

Algorithm 2: Extraction Algorithm

---

input : Stego image SI
output: Cover image $C_{(M \times N)}$, Secret image S

Algorithm Extraction():
    Step 1: Convert the stego image into **ImageArray** of red, green and blue pixels
    Step 2: Extract secret bits from every red, green and blue block of the interpolated image using the **ExtractSecretBits** function
    Step 3: Secret image is formed after extracting data from all the pixel blocks
    Step 4: The original cover image is formed by the unchanged pixels from stego image

Function ExtractSecretBits(*ImageArray*):
    Step 1: Perform the operation for each 3x3 image blocks
    Step 2: In each iteration extract secret bits from five pixel positions: $P_{i,j+1}, P_{i+1,j}, P_{i+2,j+1}, P_{i+1,j+2}$ and $P_{i+1,j+1}$ as shown in
    Fig. 2 using the **ExtractBitsFromPixel** function.

Function ExtractBitsFromPixel(*ImageArray*):
    Step 1: Extract LSB 3 from Red pixel, LSB 3 from Green pixel and LSB 2 from Blue pixel
    Step 2: Convert these data to 3 bit binary strings and append it to get 8 bit binary string
    Step 3: Convert this extracted 8 bit string to decimal data
    Step 4: Get lower bound of this decimal data and subtract it from decimal data
    Step 5: Convert this decimal data to 4 bit string which is actually the embedded 4 bit data.

---

5 Experimental Results and Comparisons

The proposed scheme has been tested using some standard benchmark images from the image database [29, 32] and evaluated using various evaluation metrics. The experimental results and comparisons are given below.

5.1 Quality Measurement Analysis

A logo image of size (330 × 330) has been used as a secret message. During the experiments, the same experimental procedure is applied to the proposed method and other methods [19, 20, 22, 25, 26, 29–31]. The performance evaluation of the proposed scheme is evaluated using PSNR, SSIM, NCC, and Q-index. The visual quality of the stego image is indicated by the PSNR. The scheme is compared with some existing PVD-based state-of-the-art schemes in terms of PSNR and embedding capacity. The values are enlisted in Table 1. Though the scheme is designed for color images, it can also be applied for the grayscale image. In that case, the 3 LSB of the interpolated pixels of the grayscale image are changed for embedding data. In Table 1, PSNR and capacity of both color image and grayscale image are given. From the table, it is observed that the average PSNR of the suggested scheme is around 44 dB, which is higher than Wu and Tsai's [12], Zhang and Wang's [30], Wang et al.'s [13], Joo et al.'s [14], Shen and Huang's [22], and Jana et al.'s [31] schemes. The maximum embedding capacity of the scheme is 2, 621, 440 bits. So,

Table 1 Comparison table of proposed scheme with other existing schemes in terms of PSNR and capacity

Schemes	Metric	Lena	Baboon	Boat	Elaine	Jet	House	Peppers	Average
Wu and Tsai [12]	Capacity	50,894	57,028	52,320	50,981	51,221	52,418	50,657	52,204
	PSNR	41.5	37	39.6	42.1	40.6	40	41.5	40.3
Zhang and Wang [30]	Capacity	50,023	53,568	50,926	49,750	50,366	51,003	49,968	50,803
	PSNR	44.3	40.4	42.4	44.5	43.5	42.7	43.9	43.1
Wang et al. [13]	Capacity	50,894	57,043	52,490	50,893	51,221	52,572	50,885	52,275
	PSNR	43.4	40.2	41.1	44.9	43.4	42.4	43.4	42.6
Joo et al. [14]	Capacity	50,894	57,043	52,490	50,893	52,662	52,572	50,815	52,275
	PSNR	43.4	39.2	41	43.5	41.5	41.5	42.5	41.9
Shen and Huang [22]	Capacity	402,485	443,472	408,777	398,250	...	412354	...	413067
	PSNR	42.46	38.88	41.6	42.98	...	41.16	...	41.42
Jana et al. [31]	Capacity	9,16,656	9,16,656	9,16,656	...	9,16,656	9,16,656	9,16,656	9,16,656
	PSNR	37.17	37.17	37.17	...	37.88	37.18	37.17	37.28
Proposed color scheme	Capacity	26,21,440	26,21,440	26,21,440	26,21,440	26,21,440	26,21,440	26,21,440	26,21,440
	PSNR	45.25	45.32	45.27	44.76	44.32	44.02	43.79	44.67
Proposed gray scheme	Capacity	9,83,040	9,83,040	9,83,040	9,83,040	9,83,040	9,83,040	9,83,040	9,83,040
	PSNR	44.37	44.82	44.19	44.09	44.17	43.87	43.18	44.09

the payload of the scheme is $= \frac{\text{Total embedded bits}}{(\text{Row} \times \text{Col})} = \frac{2621440}{(768 \times 768 \times 3)} = 1.48$ bpp, which is much higher than the existing state-of-the-art methods. The comparison graph of PSNR and embedding capacity is shown in Fig. 3. Figure 3 clarifies that, for other schemes if the PSNR is high, then the capacity is low, and if the capacity is high, then the PSNR is low. Only the proposed scheme exhibits high PSNR even when the embedding capacity is reasonably high.

The SSIM, Q-Index, and NCC values of some test images from two image data sets are shown in Table 2. SSIM is a parameter for measuring the similarity between two images. Its value approaches to +1 when two images are identical. Here, the average SSIM value is calculated around 0.9960 which indicates that the stego image is almost identical to the cover image. The average NCC and Q-index values lies between 0.9996 and 0.9999 and 0.9992 and 0.9999, respectively. These proves that their is minimum distortion in the stego image.

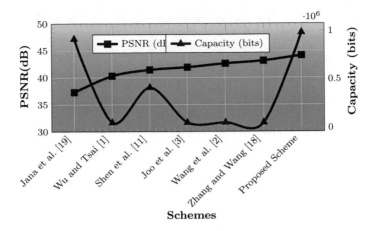

Fig. 3 Comparison graph with other existing schemes in terms of PSNR and embedding capacity

Table 2 Table for PSNR, SSIM, and NCC of different images

Image database	Cover image	PSNR	SSIM	NCC	Q-index
USC—SIPI image database (512 × 512)	Lenna	45.25	0.996	0.9999	0.99998
	Baboon	45.32	0.996	0.9998	0.99997
	Boat	45.27	0.996	0.9997	0.99996
	Elaine	44.76	0.996	0.9999	0.99992
	Jet	44.32	0.996	0.9998	0.99993
	House	44.02	0.996	0.9997	0.99994
	Peppers	43.79	0.996	0.9998	0.99995
	Soccer	44.96	0.996	0.9996	0.99996
	Zelda	43.25	0.996	0.9997	0.99997
	Barbara	44.32	0.996	0.9998	0.99998
UCID image database (512 × 512)	Ucid00008	48.3	0.996	0.9996	0.99996
	Ucid00009	47.6	0.996	0.9998	0.99997
	Ucid00011	45.3	0.996	0.9996	0.99999
	Ucid00013	48.6	0.996	0.9996	0.99998
	Ucid00015	48.3	0.996	0.9998	0.99994
	Ucid00058	47.6	0.996	0.9997	0.99995
	Ucid00061	45.3	0.996	0.9996	0.99996
	Ucid00062	48.6	0.996	0.9946	0.99993
	Ucid00128	48.6	0.996	0.9996	0.99998
	Ucid00166	44.7	0.996	0.9996	0.99994

5.2   Robustness Analysis

In this section, to verify the robustness and security of the proposed steganographic scheme, the stego image has been tested by some standard steganalysis technique like RS analysis, PVD histogram analysis, standard deviation (SD), and correlation coefficient (CC). Here the stego image is analyzed through RS analysis, and results are shown in Table 3. From the results, it is clear that the stego images successfully passed the RS analysis for $R_m \cong R_{-m}$ and $S_m \cong S_{-m}$ which proves that no stego image will not be treated as a suspicious image. So, it is very difficult for the eavesdropper to detect the secret message hidden in the stego image, which proves that the proposed scheme is secure against the RS detection attack.

PVD histogram may be a potential characteristic to expose the hidden message in the stego images generated using the PVD-based data hiding schemes. Figure 4 shows the PVD histograms of Lena image and its corresponding stego image with maximum embedding capacity. From this figure, it is observed that the PVD histograms can be well preserved after hiding secret data. Here, statistical analysis using standard deviation (SD) and correlation coefficient (CC) of the cover image (C) and stego image (SI) has been performed and the results are shown in Table 4. The SD of C and SI is 122.4311 and 122.4857, respectively, and their difference is 0.0546 for Lena image. The CC between the image C and SI is 0.9995 for Lena image which implies that it is hard to locate the embedding position within the stego image (SI). Relative entropy is the another security measurement tool. Relative entropy of the probability distribution of the cover image (C) and the stego image (SI) varies depending upon the number of bits of secret message. When the number of bits in the secret message increases, the relative entropy in stego image also increases. Relative entropy values of some standard benchmark images are listed in Table 5. From the table, it is clear

Table 3 Results of RS analysis

Image	Data (bits)	Stego image				
		R_m	R_{-m}	S_m	S_{-m}	RS value
Lena	2,621,440	59,214	52,959	39,696	45,951	0.1265
Baboon	2,621,440	36,406	33,213	25,374	28,378	0.1003
Boat	2,621,440	57,081	52,139	41,829	46,771	0.0999
Jet	2,621,440	59,807	53,035	39,103	45,875	0.1369
House	2,621,440	59,161	52,750	39,749	46,157	0.1296
Pepper	2,621,440	57,249	52,143	41,661	46,767	0.1032
Tiffany	2,621,440	56,746	62,834	42,164	36,076	0.1231
Car	2,621,440	57,945	52,682	40,965	46,228	0.1064
Barbara	2,621,440	56,955	52,302	41,955	46,608	0.0941
Zelda	2,621,440	58,990	53,727	39,920	45183	0.1064

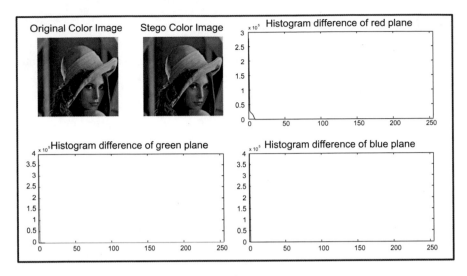

Fig. 4 Histogram difference of cover and stego image

Table 4 Standard deviation (SD) and correlation coefficient (CC) of cover image and stego image

Cover image	SD of cover image (C)	SD of stego image (SI)	CC between C and SI
Lena	122.4311	122.4857	0.9995
Baboon	134.4063	134.3275	0.9996
Boat	196.9643	196.9843	0.9998
Jet	131.7815	131.8394	0.9996
House	177.9450	177.8902	0.9998
Pepper	123.2961	123.1225	0.9995
Tiffany	77.3370	76.6433	0.9986
Car	147.8972	148.0279	0.9997
Barbara	163.2217	163.2173	0.9998
Zelda	151.4704	151.3740	0.9997

that the difference between relative entropy values of C and SI images is very small which implies that the scheme provides secure hidden communication.

5.3 Attacks

In Fig. 5, some experimental results are depicted based on some attacking conditions in stego image. Here, Tiffany image is considered for the experiment and it is tried to find out the robustness of the scheme in recovering the cover and the logo image.

Table 5 Relative entropy between original image and stego image

Image	Data (bits)	Entropy (C)	Entropy (SI)	Entropy difference
Lena	2,621,440	7.4429	7.4622	0.0193
Baboon	2,621,440	6.6752	6.7044	0.0292
Boat	2,621,440	7.2988	7.3185	0.0197
Jet	2,621,440	6.7004	6.7252	0.0248
House	2,621,440	6.4568	6.4763	0.0195
Pepper	2,621,440	6.6251	6.6464	0.0213
Tiffany	2,621,440	6.5946	6.6142	0.0196
Car	2,621,440	6.4287	6.4488	0.0201
Barbara	2,621,440	6.5746	6.5875	0.0129
Zelda	2,621,440	6.6354	6.6565	0.0211

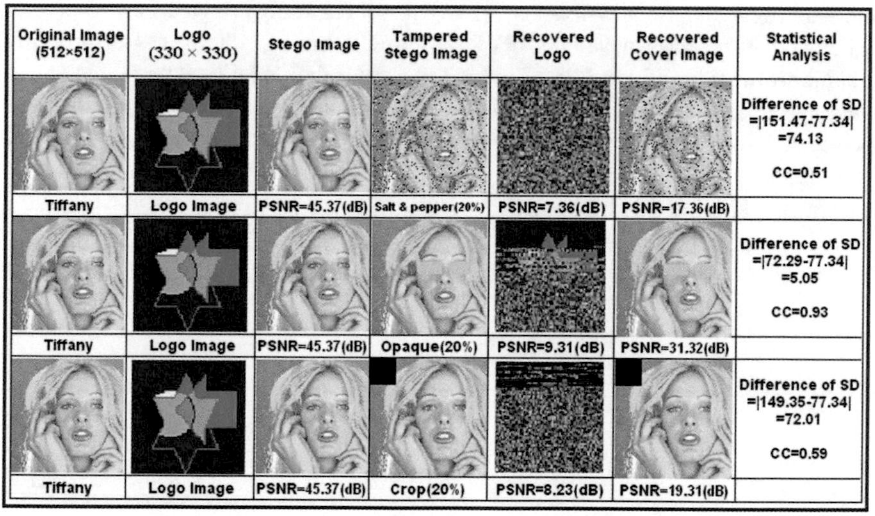

Fig. 5 Results of logo extraction from tampered stego image of proposed scheme

Three types of experiments are presented here to show the tamper detection and recovery of the original image. The first observation is salt-and-pepper noise (20% tamper) in which our scheme successfully detected that the stego image has been tampered and particular logo was not present within that tampered stego image. The difference between SD and CC is 74.13 and 0.51, respectively. The second observation is on opaque (20% tampered). In this case, it has been observed that the hidden logo has not been recovered fully during data extraction but the tampered cover image is partially recovered. The difference of SD and CC is 5.05 and 0.93, respectively. The last observation is on 20% cropping in the stego image. Here, it is also shown that the hidden logo is not recovered properly and the recovered cover image clearly shows the cropped section. The difference between SD and CC is 72.01 and 0.59, respectively. From these observations, it can be concluded that our scheme can be used for image authentication, ownership identification, copyright protection, and tamper detection.

6 Conclusion

In this approach, a reversible steganographic scheme with PVD for color image using image interpolation has been proposed. Though data hiding through PVD is generally not reversible, the reversibility has been accomplished in our approach through image interpolation. Moreover, controlling the overflow and underflow situations is an overhead to some existing schemes. We manage this by changing the LSBs instead of adding values to the existing pixel values which ensure that the overflow and underflow situations will never arise. Our scheme gains payload 6.67 bpp, that is better than recently developed PVD-based data hiding schemes. The proposed scheme still maintains good PSNR which is greater than 43.47 dB when we embed maximum secret data that is 2,621,440 bits within a (512×512) cover image. Also, the scheme gives better results to some of the measures used nowadays, like RS analysis, Chi-square analysis, and NCC, that proves the robustness of the stego images against attacking. In the future work, the scheme could be improved by using multiple pixel pairs along with multiple directions to increase the data hiding capacity as well as the imperceptibility.

References

1. Mithun, N. C., Li, J., Metze, F., & Roy-Chowdhury, A. K. (2019). Joint embeddings with multimodal cues for video-text retrieval. *International Journal of Multimedia Information Retrieval, 8*(1), 3–18.
2. Pal, P., Chowdhuri, P., & Jana, B. (2017, March). Reversible watermarking scheme using PVD-DE. In *International Conference on Computational Intelligence, Communications, and Business Analytics* (pp. 511–524). Singapore: Springer.
3. Jana B., Giri D., & Kumar, M. S. (2017). Dual-image based reversible data hiding scheme through pixel value differencing with exploiting modification direction. In *Proceedings of the First International Conference on Intelligent Computing and Communication* (Vol. 458), (pp. 549–557). Singapore: Springer.
4. Giri D., Jana B., & Kumar, M. S. (2016). Dual image based reversible data hiding scheme using three pixel value difference expansion. In *Information systems design and intelligent applications* (Vol. 434), (pp. 403-412). New Delhi: Springer.
5. Tseng, H. W., & Leng, H. S. (2013). A steganographic method based on pixel-value differencing and the perfect square number. *Journal of Applied Mathematics.*
6. Hameed, M. A., Aly, S., & Hassaballah, M. (2018). An efficient data hiding method based on adaptive directional pixel value differencing (ADPVD). *Multimedia Tools and Applications, 77*(12), 14705–14723.

7. Prasad, S., & Pal, A. K. (2017). An RGB colour image steganography scheme using overlapping block-based pixel-value differencing. *Royal Society Open Science, 4*(4), 161066.
8. Swain, G. (2018). Adaptive and non-adaptive PVD steganography using overlapped pixel blocks. *Arabian Journal for Science and Engineering, 43*(12), 7549–7562.
9. Sahu, A. K., & Swain, G. (2019). An optimal information hiding approach based on pixel value differencing and modulus function. *Wireless Personal Communications,* 1–16.
10. Luo, J., Zhou, R. G., Luo, G., Li, Y., & Liu, G. (2019). Traceable quantum steganography scheme based on pixel value differencing. *Scientific Reports, 9*(1), 1–12.
11. Swain, G. (2019). Very high capacity image steganography technique using quotient value differencing and LSB substitution. *Arabian Journal for Science and Engineering, 44*(4), 2995–3004.
12. Wu, D. C., & Tsai, W. H. (2003). A steganographic method for images by pixel-value differencing. *Pattern Recognition Letters, 24*(9–10), 1613–1626.
13. Wang, C. M., Wu, N. I., Tsai, C. S., & Hwang, M. S. (2008). A high quality steganographic method with pixel-value differencing and modulus function. *Journal of Systems and Software, 81*(1), 150–158.
14. Joo, J. C., Lee, H. Y., & Lee, H. K. (2010). Improved steganographic method preserving pixel-value differencing histogram with modulus function. *EURASIP Journal on Advances in Signal Processing, 2010*(1), 249826.
15. Liao, X., Wen, Q. Y., & Zhang, J. (2011). A steganographic method for digital images with four-pixel differencing and modified LSB substitution. *Journal of Visual Communication and Image Representation, 22*(1), 1–8.
16. Lee, Y. P., Lee, J. C., Chen, W. K., Chang, K. C., Su, J., & Chang, C. P. (2012). High-payload image hiding with quality recovery using tri-way pixel-value differencing. *Information Sciences, 191*, 214–225.
17. Liao, X., Wen, Q., & Zhang, J. (2012). A novel steganographic method with four-pixel differencing and exploiting modification direction. *IEICE Transactions on Fundamentals of Electronics, Communications and Computer Sciences, 95*(7), 1189–1192.
18. Khodaei, M., & Faez, K. (2012). New adaptive steganographic method using least-significant-bit substitution and pixel-value differencing. *IET Image Processing, 6*(6), 677–686.
19. Balasubramanian, C., Selvakumar, S., & Geetha, S. (2014). High payload image steganography with reduced distortion using octary pixel pairing scheme. *Multimedia Tools and Applications, 73*(3), 2223–2245.
20. Chen, J. (2014). A PVD-based data hiding method with histogram preserving using pixel pair matching. *Signal Processing: Image Communication, 29*(3), 375–384.
21. Jung, K. H., & Yoo, K. Y. (2015). High-capacity index based data hiding method. *Multimedia Tools and Applications, 74*(6), 2179–2193.
22. Shen, S. Y., & Huang, L. H. (2015). A data hiding scheme using pixel value differencing and improving exploiting modification directions. *Computers & Security, 48*, 131–141.
23. Swain, G. (2016). Adaptive pixel value differencing steganography using both vertical and horizontal edges. *Multimedia Tools and Applications, 75*(21), 13541–13556.
24. Lutovac, B., Daković, M., Stanković, S., & Orović, I. (2017). An algorithm for robust image watermarking based on the DCT and Zernike moments. *Multimedia Tools and Applications, 76*(22), 23333–23352.
25. Hussain, M., Wahab, A. W. A., Ho, A. T., Javed, N., & Jung, K. H. (2017). A data hiding scheme using parity-bit pixel value differencing and improved rightmost digit replacement. *Signal Processing: Image Communication, 50*, 44–57.
26. Jung, K. H. (2018). Data hiding scheme improving embedding capacity using mixed PVD and LSB on bit plane. *Journal of Real-Time Image Processing, 14*(1), 127–136.
27. Dorfer, M., Schlüter, J., Vall, A., Korzeniowski, F., & Widmer, G. (2018). End-to-end cross-modality retrieval with CCA projections and pairwise ranking loss. *International Journal of Multimedia Information Retrieval, 7*(2), 117–128.
28. Nagai, Y., Uchida, Y., Sakazawa, S., & Satoh, S. I. (2018). Digital watermarking for deep neural networks. *International Journal of Multimedia Information Retrieval, 7*(1), 3–16.

29. *UCID image database*. UK: Nottingham Trent University. http://jasoncantarella.com/downloads/ucid.v2.tar.gz.
30. Zhang, X., & Wang, S. (2006). Efficient steganographic embedding by exploiting modification direction. *IEEE Communications Letters, 10*(11).
31. Jana, B., Giri, D., & Mondal, S. K. (2016). Dual-image based reversible data hiding scheme using pixel value difference expansion. *IJ Network Security, 18*(4), 633–643.
32. *The USC-SIPI image database*. University of Southern California. http://sipi.usc.edu/database/database.php.

Lattice-Based Fuzzy Medical Expert System for Management of Low Back Pain: A Preliminary Design

Debarpita Santra, S. K. Basu, J. K. Mandal, and Subrata Goswami

Abstract Low back pain (LBP) appears to be the foremost contributor to the years lived with disability globally, depriving many individuals across the nations of leading daily activities. Diagnosing LBP is quite challenging as it requires dealing with several clinical variables having no precisely quantified values. With the goal to design a reliable medical expert system for assessment and management of LBP, the research offers a lattice-based scheme for efficient representation of relevant medical knowledge, proposes a suitable methodology for design of a fuzzy knowledge base that can handle imprecision in knowledge, and derives a fuzzy inference system. A modular approach is taken to construct the fuzzy knowledge base, where each module is able to capture interrelated clinical knowledge about medical history, findings of physical examinations, and pathological investigation reports. The fuzzy inference system is designed based on the Mamdani method. For fuzzification, the design adopts triangular membership function; for defuzzification, the centroid of area technique is used. With the relevant medical knowledge being acquired from the expert physicians, a working prototype of the system has been built. The prototype has been successfully tested with some LBP patient records available at the ESI Hospital, Sealdah. The designed prototype is found to be clinically acceptable among the expert and non-expert physicians.

D. Santra (✉) · J. K. Mandal
Department of Computer Science and Engineering, University of Kalyani, Kalyani, Nadia, West Bengal 741235, India
e-mail: debarpita.cs@gmail.com

J. K. Mandal
e-mail: jkm.cse@gmail.com

S. K. Basu
Department of Computer Science, Banaras Hindu University, Varanasi 221005, India
e-mail: swapankb@gmail.com

S. Goswami
ESI Institute of Pain Management, Kolkata 700009, India
e-mail: drsgoswami@gmail.com

© Springer Nature Singapore Pte Ltd. 2020
J. K. Mandal and S. Banerjee (eds.), *Intelligent Computing: Image Processing Based Applications*, Advances in Intelligent Systems and Computing 1157,
https://doi.org/10.1007/978-981-15-4288-6_8

119

Keywords Medical expert system · Low back pain (LBP) · Fuzzy logic · Lattice
theory · Knowledge base

1 Introduction

India is confronting a huge void in its healthcare infrastructure, which can be filled
with enormous application of information technology in the existing solutions. Medical expert system [1], a computer-based new-generation medical decision support
tool powered by artificial intelligence (AI), bridges the gap between medical and
computer sciences by engaging doctors and computer scientists to work in a collaborative manner so that the clinical parameters, symptoms, knowledge of the expert
doctors can be integrated in one system.

This paper deals with developing a medical expert system for management of
low back pain (LBP) [2]. LBP, which is a musculoskeletal disorder, affects huge
population world-wide belonging to both the affluent and the poor socio-economic
strata of society. With no standard LBP diagnosis and management guideline being
followed in many countries like India, assessment or evaluation of LBP needs detailed
knowledge about human anatomy and physiology. With very few standard techniques
accepted and widely used in clinical practice for measuring pain accurately, it is
quite difficult to describe pain in terms of crisp quantifiable attribute values. Also,
how a person feels pain depends on his/her physiological, psychological, or other
environmental factors. That is why the pain intensity of an individual may differ
from the other. These kinds of clinical uncertainties can be resolved with the help of
fuzzy logic and other mathematical techniques.

While fuzzy logic deals with imprecision in knowledge, the properties like completeness and non-redundancy in the knowledge base are ensured by the application
of lattice theory in the design. This paper focuses on the design of a lattice-based
fuzzy medical expert system for diagnosis of LBP. The triangular membership functions have been used for fuzzifying the linguistic values which are used to describe
the clinical attributes, and the Mamdani inference approach has been used for designing the inference engine. The system has been successfully tested with twenty LBP
patient records available at the ESI Hospital Sealdah, West Bengal, India.

The paper is oriented as follows: Sect. 2 provides an outline of well-known medical
expert systems as well as existing fuzzy-logic-based design approaches. Sect. 3
addressed the design issues of the intended medical expert system. In Sect. 4, the
proposed system has been illustrated with a small example adopted from the domain
of LBP. At last, conclusion of the paper is drawn in Sect. 5.

2 Previous Works

Development of medical expert systems began in early 70's with MYCIN [3], followed by CASNET [4], INTERNIST [5], ONCOCIN [6], PUFF [7], etc. As uncertainty pervades in almost all stages of clinical decision making, some existing medical expert systems used different mathematical techniques to handle inconsistency, imprecision, incompleteness, or other uncertainty-related issues. While rough set theory [8] is widely used for addressing mainly incompleteness or inconsistency in knowledge, fuzzy logic [9] is extensively used for handling imprecision in knowledge. A number of fuzzy medical expert systems have been reported in the literature for diagnosis of heart disease, typhoid fever, abdominal pain, and for providing clinical support in intensive care unit. A brief outline of the fuzzy-based developments is provided in this section.

A fuzzy-logic-based medical expert system for diagnosis of heart disease was proposed based on the databases available in Cleveland Clinic, V.A. Medical Centre, and Long Beach databases. The system took help of Mamdani inference method with thirteen input parameters, and single output parameter [10]. The decision support system that aids in diagnosing typhoid fever used triangular membership function for each of the decision variables, and also applied Mamdani fuzzy inference method for designing the inference system. The proposed system was successfully evaluated with the typhoid patient cases from Federal Medical Centre, Owo, Ondo State, Nigeria [11]. For diagnosis of acute abdominal pain, a fuzzy-logic-based expert system was proposed in the literature that contained around two hundred symptoms. Knowledge about these symptoms was stored in the knowledge base in the form of around sixty rules and four thousand fuzzy relations. The system was partially tested with hundred patients at the 'Chirurgische Klinik' in Dusseldorf [12].

A fuzzy medical expert system was designed in [13] for diagnosis of back-pain-related diseases originating from various spinal regions (cervical, thoracic, lumbar, sacral, and coccygeal). The design considered few clinical parameters like patient's body mass index, age, gender, and history of other disease(s).

The above discussions offer a background about how AI tools and fuzzy algorithms have been used for assisting physicians through development of medical expert systems in different medical domains like meningitis, glaucoma, pulmonary disease, oncology, skin disease, ectopic pregnancy, heart disease, and so on. But the domain of LBP has not been under focus of research over years. The application of AI for assessment and management of LBP is in emergent stage till now. The AI-based approaches discussed above are not directly applicable for LBP domain as this domain has different kinds of complex inter-dependencies among the clinical parameters. The paper proposes a preliminary design for a fuzzy-based development of a medical expert system to aid clinically mainly junior doctors, general physicians, or other healthcare providers in especially remote clinical settings to provide fast, quality, and affordable treatment for the domain of LBP.

3 Design of Proposed System

An efficient development of the intended system requires careful design of mainly four components: user interface, fuzzy knowledge base, working memory, and fuzzy inference system, as shown in Fig. 1. The user interface acts as a gateway between the pertinent users (attending physicians), and the system itself. The knowledge base stores vast amount of updated and exhaustive medical knowledge acquired from the existing literature involving international guidelines on LBP management, relevant review articles or other journal papers, randomized control trials, case studies, cohort studies, etc., and also from the wisdom gleaned from expert physicians in this domain. The imprecision in knowledge is handled through use of fuzzy logic. The fuzzy inference system is basically the decision-making unit, which accepts patient's information inputted through the user interface, fuzzifies the input with the help of a fuzzifier, processes the input to offer diagnostic conclusions at the level of expert physicians with the help of a fuzzy inference engine, and finally, defuzzifies the inference outcomes with the help of a defuzzifier to reflect the final results to the user interface. The working memory holds temporary results generated during the inference process.

When an LBP patient comes for treatment, all the significant clinical information (clinical history, reports of clinical examinations, results of pathological investigations, etc.) of the patient are fed into the system through the user interface. The fuzzy inference system accepts the inputted patient information, matches them with the knowledge stored in knowledge base, and outputs a probable list of diseases from which the patient may be suffering from. The following subsections describe how the knowledge base and the inference engine are constructed for the intended design.

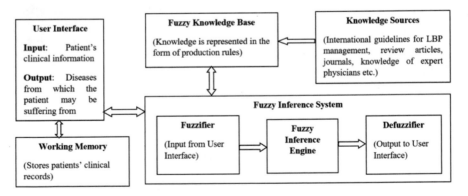

Fig. 1 Block diagram of the intended fuzzy medical expert system for LBP diagnosis

3.1 Design of Fuzzy Knowledge Base

We assume that there is a finite set D of x (> 0) LBP diseases (classified according to different pain generators such as joints or bones, intervertebral discs, nerve roots, muscles, etc., [14]) in the literature. Assessment of different LBP diseases goes through three different phases: *phase* 1 for finding a list of possible LBP disorders after careful consideration of relevant clinical history of a patient X; *phase* 2 for determining a list of probable diseases based on the clinical examinations of X; and *phase* 3 for finding a list of likely LBP diseases evaluated through reviewing the clinical investigation results. Considering that there is a finite nonempty set A of n clinical parameters/attributes relevant for assessment of all the LBP diseases in the literature, n_1 ($0 < n_1 < n$) clinical attributes comprising the set A_1 ($\subset A$) are used for collection of clinical history in *phase* 1, n_2 ($0 < n_2 < n$) attributes forming the set A_2 ($\subset A$) are there for acquiring the clinical examination reports of LBP patients in *phase* 2, and n_3 ($0 < n_3 < n$) attributes constituting the set A_3 ($\subset A$) are considered for capturing the clinical investigation results of LBP patients in *phase* 3. Here, A_1, A_2, A_3 are disjoint subsets of A.

With no standard techniques being accepted and widely-used in clinical practice for measuring pain accurately, it is quite difficult to describe pain in terms of crisp quantifiable attribute values. Use of numeric rating scale (NRS) as shown in Fig. 2 for measuring pain intensity is a popular approach.

In NRS, the three pain intensity values 'No Pain', 'Moderate Pain', and 'Severe Pain' cannot be crisply defined; rather, there exists imprecision in defining the range for every linguistic value.

The fuzzy knowledge base for the proposed medical expert system holds knowledge related to x LBP diseases, where each disease is characterized by the clinical attributes belonging to A. Every piece of knowledge needs to be represented in the form of production rules, which are constructed from the acquired knowledge in a systematic fashion with the help of a lattice structure. The lattice structure actually works as the fuzzy knowledge base.

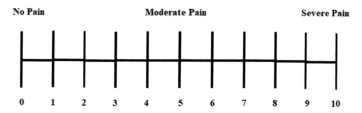

Fig. 2 Numeric rating scale for pain assessment

3.1.1   Rule Generation Using Lattice

Prior to formation of a lattice structure, a fuzzy information system S_f is designed using the acquired knowledge, which contains M (> 0) rows and N ($= n + 2$) columns. In S_f, each tuple is represented as $< (d, v_f(d)), \{(a_j, v_f(a_j)) \mid a_j \in A \ (1 \leq j \leq n)$ is the j-th clinical attribute for the disease d, and $v_f(a_j)$ denotes the linguistic value of $a_j\}, r_s >$, where d represents an LBP disease, $v_f(d)$ represents the linguistic value associated with the chance of occurrence of d, and r_s denotes the reliability strength of a production rule of the form $[\cup_{j=1 \text{ to } n} \{(a_j, v_f(a_j))\} \rightarrow (d, v_f(d))]$. The reliability strength r_s ($0 \leq r_s \leq 1$), which measures how much reliable the piece of knowledge is, is determined based on the nature of knowledge sources used for acquiring the knowledge. For example, if a piece of knowledge is acquired from an international guideline, the reliability strength for the piece of knowledge is very high, compared to that of another piece of knowledge acquired from a case study or expert consensus.

The fuzzy information system S_f can formally be represented as $S_f = (D_f, A_f, R_f)$, where D_f is a set of (disease, linguistic_value) pairs, A_f is a set of (attribute, linguistic_value) pairs, and R_f is the set of corresponding reliability strengths. For every subset B_f of A_f, a binary indiscernibility relation $I(B_f)$ [8] on D_f is defined as follows: two different (decision, linguistic_value) pairs $(d_i, v_f(d_i))$ and $(d_j, v_f(d_j))$ [$i \neq j$] are indiscernible by the subset of attributes B_f in A_f, if the linguistic value corresponding to an attribute $b_f \in B_f$ is the same for both the pairs $(d_i, v_f(d_i))$ and $(d_j, v_f(d_j))$. The equivalence class of the relation $I(B_f)$ is called an elementary set in B_f [8]. The construction of equivalence classes involving the elements in A_f and D_f is performed in a systematic fashion, as given below in a procedure called *Equivalence_Class_Construction*(). The algorithm uses the fuzzy information system S_f as input.

Procedure *Equivalence_Class_Construction* (S_f)
Begin
   $k_1 = 0$ // k_1 is a temporary variable
   $E_f = \varnothing$ // E_f is the set of elementary sets, which is initially empty
   While $k_1 \leq n$ do //n is the no. of clinical attributes
      For $k_2 = 1$ to $^nC_{k1}$ do   // k_2 is a temporary variable
      $B_f = $ ***Combination***(A_f, k_1) // The procedure ***Combination***(A_f, k_1) returns a
                                set of k_1 clinical attributes from a set of n
                                attributes in each iteration
      $E_f = E_f \cup [D_f/B_f]$ // $[D_f/B_f]$ denotes the elementary sets of D_f determined by B_f
      End for
      $k_1 = k_1 + 1$
   End while
   Return E_f
End *Equivalence_Class_Construction*

If a maximum of K instructions is executed by the procedure ***Combination***() in ***Equivalence_Class_Construction***() procedure, the total number of operations performed by the latter is $N_e = (\sum_{k_1=0 \text{ to } n} K \times k_1 \times {}^n C_{k_1})$. The algorithm has time and space complexities as $O(N_e)$ and $O(E_f)$, respectively.

All the knowledge contained in the fuzzy information system S_f is stored efficiently in the fuzzy knowledge base KB_f. Initially, $KB_f = \emptyset$. Every time, a new clinical attribute is encountered, information regarding the attribute is added to the existing KB_f using 'set union' operation. With n (> 0) clinical attributes in A, the knowledge base is modified as $KB_f = KB_f \cup \{a_i \mid a_i \ (\in A)$ is a clinical attribute, where $1 \leq i \leq n\}$. To extract detailed knowledge about each of the n elements in KB_f, the attributes are combined with each other in an orderly fashion. First, ${}^n C_2$ combinations are made taking exactly two different attributes at a time from the n attributes. In this case, $KB_f = KB_f \cup \{(a_i, a_j) \mid a_i$ and a_j are the two clinical attributes in A, where $a_i \neq a_j$ and $1 \leq i, j \leq n\}$. In the same way, ${}^n C_k$ combinations of attributes are found, where $2 \leq k \leq n$, and the knowledge base is updated accordingly. Now, the KB_f holds total of ($2^n - 1$) attribute combinations. So, a set S constructed using all the attribute combinations in KB_f acts as the power set of A and is represented as $S = \{\phi, \{a_1\}, \{a_2\}, \{a_3\}, ..., \{a_n\}, \{a_1, a_2\}, \{a_1, a_3\}, ..., \{a_{n-1}, a_n\}, \{a_1, a_2, a_3\}, ..., \{a_{n-2}, a_{n-1}, a_n\}, ..., \{a_1, a_2, ..., a_{n-1}, a_n\}\}$.

Now, a subset equality ($\subseteq$) relation R is considered. Assume that $\alpha, \beta, \gamma \subseteq S$. So, [$S$; $\subseteq$] can be called as a poset because it holds reflexivity ($\forall \alpha \in S : (\alpha \subseteq \alpha)$), antisymmetry ($\forall (\alpha, \beta) \in S : \alpha \subseteq \beta$ and $\beta \subseteq \alpha \Rightarrow (\alpha = \beta)$), and transitivity ($\forall (\alpha, \beta, \gamma) \in S : \alpha \subseteq \beta, \beta \subseteq \gamma \Rightarrow (\alpha \subseteq \gamma)$) relationships.

Now, a set $\{x', y'\}$ is formed with two distinct elements x' and y' in set S. A greatest lower bound (glb) as $(x' \cap y')(\in S)$ and a least upper bound (lub) as $(x' \cup y')(\in S)$ would always exist in this case. So, S is termed as a lattice having order n. Each node in the lattice contains three types of information: a set B_f of clinical attributes associated with their corresponding linguistic values as recorded in S_f, the elementary sets obtained against B_f using the procedure *Equivalence_Class_Construction*(), and the reliability strength against each element belonging to the elementary sets.

Suppose, a node in the lattice S contains m (> 0) elementary sets $E_1, E_2, ..., E_m$ against the set of attributes B_f. Assume that B_f contains k (> 0) clinical attributes $\{a_i, a_{i+1}, a_{i+2}, ..., a_{i+k}\}$ with each attribute a_{k1} ($i \leq k1 \leq (i + k)$) of B_f has a linguistic value v_{k1}. The corresponding elementary set E_j ($1 \leq j \leq m$) contains l (> 0) (d, $v_f(d)$) pairs, where d denotes an LBP disease and $v_f(d)$ denotes the linguistic value associated with the chance of occurrence of the disease d. A maximum of l production rules would be generated of the form $[(a_i, v_i)$ AND (a_{i+1}, v_{i+1}) AND ... AND $(a_{i+k}, v_{i+k}) \rightarrow E_j]$. If all the l ($\leq x$) elements in E_j contains l different LBP diseases, then the total number of production rules generated against B_f is l. If on the other hand, at least two elements in E_j contain the same disease with two distinct linguistic values val_1 and val_2 respectively, the issue is resolved with the help of associated reliability strengths. An item with greater reliability strength obviously gets priority over the other elements. The same reliability strength for more than one such conflicting items leads to incorporation of another mathematical technique(s) for handling inconsistency in knowledge. Handling of inconsistencies is out of scope for this paper.

3.2 Design of Fuzzy Inference System

The fuzzy-logic-based inference system, as shown in Fig. 1, consists of a fuzzifier, a defuzzifier, and an inference engine. The functionality of the fuzzifier is to fuzzify the raw input which contains basically the linguistic values associated with the clinical attributes belonging to A using triangular membership function. Corresponding to the set A, a fuzzy set A_F is first defined as $A_F = \{(a_i, \mu(a_i)) | a_i \in A, \mu(a_i) \in [0, 1]\}$, where $\mu(a_i)$ denotes the membership function of a_i.

Any inference engine of a medical expert system performs two tasks: (i) it matches the inputted patient information with the stored knowledge; (ii) the matched knowledge is processed for making reliable diagnostic conclusions. Use of a lattice structure as the knowledge base makes the matching process easier. As the information about an LBP patient may include only p ($\leq n$) clinical attributes, it would be sufficient only to search the nodes at the p-th level of the lattice S. This kind of matching strategy reduces the search time in the knowledge base to a great extent. The fuzzy inference engine uses the Mamdani approach [11], and the defuzzification process uses the centroid of area method [11].

As LBP diagnosis goes mainly through three phases, three different lattice structures are formed to store knowledge. The inference engine is also executed phasewise, and the three different lists of probable diseases are obtained from these phases. In each phase, the list of probable diseases gets refined. The list of diseases obtained after *phase* 3 would be regarded as the final outcomes of the medical expert system. This type of modular approach ensures scalability in the design.

4   Illustration and Implementation

For the sake of simplicity, this paper considers only five important clinical history parameters namely 'pain at low back area' (a_1), 'pain at legs' (a_2), 'pain at rest' (a_3), 'pain during forward bending' (a_4), and 'pain during backward bending' (a_5) as input. No clinical examination and investigation parameters have been used in this paper for implementation. Only three linguistic values 'No', 'Moderate', and 'Severe' are considered as the values for the set A_1 of attributes $\{a_1, a_2, a_3, a_4, a_5\}$, and the ranges of these fuzzy values are defined as [0, 4], [3, 7], and [6, 10], respectively. The membership function graph for these linguistic variables is shown in Fig. 3a.

The designed medical expert system outputs the chances of occurrence for only five major LBP diseases, namely 'Sacroiliac Joint Arthropathy' (d_1), 'Facet Joint Arthropathy' (d_2), 'Discogenic Pain' (d_3), 'Prolapsed Intervertebral Disc Disease' (d_4), and 'Myofascial Pain Syndrome' (d_5). Chance of occurrence of every disease can be represented through four linguistic values 'No', 'Low', 'Moderate', and 'High', with the respective ranges of values being defined as [0, 10], [8, 25], [20,

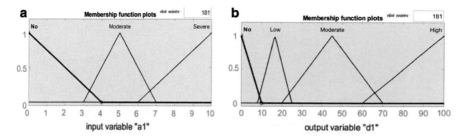

Fig. 3 **a** Membership function for a_1, **b** Membership function for d_1

70], and [60, 100], respectively. The membership function graph corresponding to the chance of occurrence of the LBP diseases is shown in Fig. 3b.

Using the acquired knowledge, an information system S_f is first constructed involving the clinical attributes in A_1, as shown in Table 1. Using the attributes in A_1, a lattice structure is constructed as shown in Fig. 4. Against each node

Table 1 Information system S_f corresponding to clinical attributes in A_1

(Disease, linguistic value)	a_1	a_2	a_3	a_4	a_5	Reliability strength (r_s)
(d_1, High)	Moderate	No	Severe	Moderate	Moderate	0.8
(d_2, High)	Moderate	No	No	No	Severe	0.7
(d_3, Moderate)	Moderate	No	No	No	No	0.6
(d_4, High)	No	Severe	No	Severe	No	0.6
(d_5, Low)	Moderate	No	Moderate	No	No	0.4

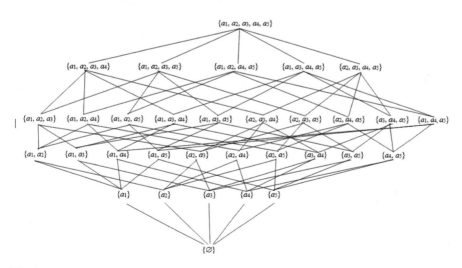

Fig. 4 Lattice structure formed using the attributes in A_1

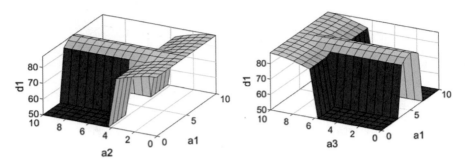

Fig. 5 Surface viewers for disease d_1 against the attributes a_1, a_2, and a_3

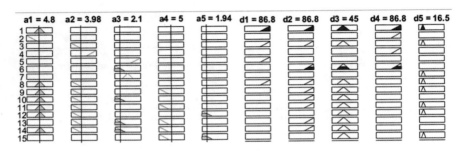

Fig. 6 Partial snapshot of the rule-viewer for the sample test case

in the lattice, the elementary sets are obtained using the procedure ***Equivalence_Class_Construction***() and the information system S_f. In this way, the fuzzy knowledge base is constructed.

For the proposed fuzzy expert system, the surface viewers of some fields are shown in Fig. 5.

A prototype of the designed system has been tested with 20 LBP patient cases obtained from the ESI Hospital, Sealdah, West Bengal. No standard systematic benchmark datasets are publicly available for proper evaluation of the proposed methodology. For 18 cases, among the considered dataset, the achieved results have matched with the expected results. A sample test case has been taken with input values for $a_1 = 4.8$, $a_2 = 3.98$, $a_3 = 2.1$, $a_4 = 5$, and $a_5 = 1.94$. The partial-snapshot of the rule-viewer for this scenario is shown in Fig. 6. The outputs from the rule-viewer say that the chance of occurrence of the diseases d_1, d_2, and d_4 are high, while that of d_3 is moderate, and that of d_5 is low.

5 Conclusion

The advantages of developing a lattice-based fuzzy medical expert system are system completeness, preciseness, and zero-redundancy in the knowledge base. The proposed design methodology for construction of knowledge base and derivation of fuzzy inference strategy act as a firm basis for development of the intended expert system for LBP management. As the number of LBP diseases is upper bounded by 10 as per expert knowledge, the time and space complexity would not be of much concern. The proposed schemes can be easily extended in near future for design of a full-fledged medical expert system for assessment and management of LBP management. The system would offer a fast, quality, and affordable medical solution for this ailment to large cohort of patients suffering from LBP.

Acknowledgements The authors are thankful to the ESI Hospital Sealdah, Kolkata, India, for providing access to LBP patient records for fulfilment of this research work.

References

1. Liao, S. H. (2005). Expert system methodologies and applications—A decade review from 1995 to 2004. *Expert Systems with Applications, 28*(1), 93–103.
2. Andersson, G. B. (1999). Epidemiological features of chronic low-back pain. *The Lancet, 354*(9178), 581–585.
3. Shortliffe, E. (Ed.). (2012). *Computer-based medical consultations: MYCIN* (Vol. 2). Elsevier.
4. Kulikowski, C. A., & Weiss, S. M. (1982). Representation of expert knowledge for consultation: The CASNET and EXPERT projects. *Artificial Intelligence in Medicine, 51*, 21–55.
5. Miller, R. A., Pople, H. E., Jr., & Myers, J. D. (1982). Internist-I, an experimental computer-based diagnostic consultant for general internal medicine. *New England Journal of Medicine, 307*(8), 468–476.
6. Shortliffe, E. H., Scott, A. C., Bischoff, M. B., Campbell, A. B., Van Melle, W., & Jacobs, C. D. (1984). An expert system for oncology protocol management. In Buchanan, B. G. and Shortiffe, E. H. (Eds.), *Rule-based expert systems* (pp. 653–665).
7. Aikins, J. S., Kunz, J. C., Shortliffe, E. H., & Fallat, R. J. (1983). PUFF: An expert system for interpretation of pulmonary function data. *Computers and Biomedical Research, 16*(3), 199–208.
8. Pawlak, Z. (1998). Rough set theory and its applications to data analysis. *Cybernetics and Systems, 29*(7), 661–688.
9. Klir, G. J., & Yuan, B. (1995). *Fuzzy sets and fuzzy logic: Theory and applications.* Upper Saddle River (p. 563).
10. Adeli, A., & Neshat, M. (2010). A fuzzy expert system for heart disease diagnosis. In *Proceedings of international multi conference of engineers and computer scientists, Hong Kong* (Vol. 1).
11. Samuel, O. W., Omisore, M. O., & Ojokoh, B. A. (2013). A web based decision support system driven by fuzzy logic for the diagnosis of typhoid fever. *Expert Systems with Applications, 40*(10), 4164–4171.
12. Fathi-Torbaghan, M., & Meyer, D. (1994). MEDUSA: A fuzzy expert system for medical diagnosis of acute abdominal pain. *Methods of Information in Medicine, 33*(5), 522–529.
13. Kadhim, M. A., Alam, M. A., & Kaur, H. (2011). Design and implementation of fuzzy expert system for back pain diagnosis. *International Journal of Innovative Technology & Creative Engineering, 1*(9), 16–22.
14. Allegri, M., Montella, S., Salici, F., Valente, A., Marchesini, M., Compagnone, C., and Fanelli, G. (2016). Mechanisms of low back pain: A guide for diagnosis and therapy. *F1000Research, 5.*

An Optimized Intelligent Dermatologic Disease Classification Framework Based on IoT

Shouvik Chakraborty, Sankhadeep Chatterjee, and Kalyani Mali

Abstract Internet of things (IoT) is one of the recent concepts that provide many services by exploiting the computational power of several devices. One of the emerging applications of the IoT-based technologies can be observed in the field of automated health care and diagnostics. With the help of IoT-based infrastructures, continuous data collection and monitoring are simpler. In most of the scenarios, collected data are massive, unstructured, and contain many redundant parts. It is always a challenging task to find an intelligent way to mine some useful information from a massive dataset with stipulated computing resources. Different types of sensors can be used to acquire data in real time. Some sensors can be body-worn sensors, and some sensors can be placed some distance apart from the body. In dermatological disease detection and classification problem, images of the infected region play a vital role. In this work, an optimized classification method is proposed that can be useful in performing automated classification job in the limited infrastructure of the IoT environment. The input features are optimized in such a way so that it can be useful in faster and accurate classification by the classifier that makes the system intelligent and optimized. Moreover, the hybrid classifier is optimized using different metaheuristic optimization methods for better convergence. The proposed work can be highly beneficial in exploring and applying the power of IoT in the healthcare industry. It is a small step toward the next-generation healthcare systems which can produce faster and accurate results at affordable cost with the help of IoT and remote healthcare monitoring.

S. Chakraborty · K. Mali
Department of Computer Science and Engineering, University of Kalyani, Kalyani, India
e-mail: shouvikchakraborty51@gmail.com

K. Mali
e-mail: kalyanimali1992@gmail.com

S. Chatterjee (✉)
Department of Computer Science and Engineering, University of Engineering and Management,
Kolkata 700160, India
e-mail: chatterjeesankhadeep.cu@gmail.com

© Springer Nature Singapore Pte Ltd. 2020
J. K. Mandal and S. Banerjee (eds.), *Intelligent Computing: Image Processing
Based Applications*, Advances in Intelligent Systems and Computing 1157,
https://doi.org/10.1007/978-981-15-4288-6_9

Keywords Artificial intelligence · Artificial neural networks · Biomedical image analysis · Internet of things · Remote health care

1 Introduction

Recent developments in technology are providing elegant solutions to many problems. Due to advent of sophisticated sensors and other wearable devices, data collection and monitoring become cheaper and easier. Physiological information can be recorded continuously which can be useful in tracking the status of the patient. It is inexpensive and helpful in routine monitoring of health. IoT-based infrastructures can provide a convenient way to collect different parameters from various regions of a patient and transmit it in real time. It can be helpful for the doctors because it can record and store dynamic data and does not depend only on the static measurements of the laboratories. The data collected using IoT-based devices not only provide an efficient mechanism for data collection but it can also mine some important information related to health care. It is possible to obtain significant information from the collected data by real-time continuous data analysis. Doctors can provide more accurate advice with the help of the recorded data collected in real time [1]. Not only medicines, doctors can also suggest the appropriate daily routine that is suitable for a specific patient. Moreover, doctors can design some customized medicines depending on the different parameters and lifestyle of a particular patient that can accelerate the recovery process. So, efficient use of the IoT-based healthcare systems can improve the health of the current healthcare industry as well as technical advancement can be achieved [2]. The cost and time can be significantly reduced with increased accuracy and belief, which can change the global view for the healthcare industry [3].

IoT can bring revolution in different fields of science and technology in the near future. Different sensors are available which can be used to record various physical parameters [4]. There are many problems associated with the IoT-based systems. First of all, some redundant parts can be present inside the data which are collected by the sensors and in most of the cases, the collected data are unstructured [5]. These data incur more transmission overhead (in terms of power consumption). Moreover, it is very difficult to process such a large amount of data (containing some insignificant part) and get accurate results in real time using optimal power. So, technological improvement is only possible if sophisticated data structures and algorithms can be devised that use less computational resources and capable to produce accurate results in stipulated amount of time [6, 7]. In IoT-based environment, energy cannot be wasted to deal with the redundant data. Moreover, efficient methods are required to extract desired information from the data collected by the sensors. In case of biomedical image data, the processing algorithms must be very efficient so that it can extract the result within reasonable time and with high accuracy against the minimal power, otherwise large-scale implementation of such systems is impossible. In this work, the biomedical image analysis and machine-learning domains are exploited to integrate the modern healthcare industry with the benefits of computer-aided diagnostic [8]. The proposed framework integrates the machine-learning methods with IoT for efficient processing of the collected data that will be beneficial for large-scale implementation of the IoT-based systems for biomedical data analysis. In general, the size of the image data is quite large and therefore, processing such data in IoT environment is always challenging. This work tries to reduce the redundant data and an automated classification framework is proposed that perform the job within the limited infrastructure of the IoT environment. The proposed framework is intelligent and optimized due to its inherent capability of optimizing the feature space and the convergence of the hybrid neural classifier. The input features are optimized in such a way so that it can be useful in faster and accurate classification by the classifier. Moreover, the hybrid classifier is optimized using the different metaheuristic optimization methods for better convergence. It is a small step toward the large-scale implementation of the next-generation healthcare systems that can produce faster and accurate results in affordable cost with the help of IoT.

The rest of the paper is organized as follows. Section 2 gives an overview of the task with some background information. Section 3 gives illustration of the system architecture of the machine-learning-coupled IoT-based healthcare system. Section 4 describes the proposed skin lesion analysis and classification method. Section 5 describes and interprets the obtained results. Section 6 concludes the article with some future research possibilities in this domain based on the current work.

2 Background and Overview

The very first step of any IoT-based remote healthcare and monitoring system is to capture information from the part of the body or different parameters from the whole body. The acquired data is then communicated over the network and the processing units process the data with the limited computational power [9]. It can exploit the processing power of several devices which are connected in the IoT architecture. In [9], authors developed an IoT-based framework where different body parameters are recorded using some wearable sensors using wireless body area network (WBAN) [10, 11]. The Bluetooth technology was used to transmit data to some gateway servers. From this server, the data is transferred to another remote server. In [10], authors propose a dermoscopic disease classification framework for machine learning. A skin disease detection and classification methods are proposed in [11]. An automated diagnosis method to diagnose skin carcinoma is proposed in [12]. This method is based on the parallel deep residual networks. A mutual learning-based method is proposed in [13] for skin lesion classification. A deep-learning-based method is proposed and several deep-learning-based methods are analyzed in [14] for melanoma classification. In [15], authors proposed a new learning method to classify cross-domain skin diseases. A feature extraction method is proposed in [16] for skin lesions classification from dermoscopic images. Physicians can access the data and can track the values of different parameters. Another approach is presented in [12]. Here, retrieved data are stored and processed in a cloud-based environment. Physicians can access and view the data along with processed information through online application. Both automated processing and manual processing are possible in the cloud-based environment. Different powerful sensors are available at an affordable cost which makes the data acquisition easier [17]. But the main job is to process and mine meaningful information from the collected data because raw data is not useful in most of the cases. So, the data analysis module is one of the most vital components of any IoT-based healthcare system [13]. The collected data is often found to be high dimensional. Moreover, a huge amount of data can be recorded by these sensors. Hence, it is very difficult for the physicians or other experts to analyze these data manually. Moreover, manual data analysis can introduce some error in the results accidentally and some hidden information may not be revealed. So, an efficient automated processing framework is the heart of the remote healthcare system. Otherwise, the collected data cannot be used effectively. Accurate and timely diagnosis relies on the appropriate data visualization [14] and mining. At the inception phase of IoT-based healthcare systems, the major concern was concentrated on the transmission and storage of the data. Some recent researches show that the one of the major part of current focuses is on the processing part of the remote healthcare systems [15].

One of the major advantages of the IoT-based devices is their availability. Information can be captured and exchanged at any point of time through IoT-based infrastructure. It increases the reliability, flexibility, and processing power. There are several advantages of being connected. The supervising device can send information and alarm signal to the nearest hospital of a patient in case of any emergency. It is helpful for many patients and can save many precious lives [16]. So, the IoT-based healthcare systems with appropriate technological support can convert this concept into blessings for many people.

In this work, an IoT-based dermoscopic machine-learning-based image analysis method is developed that can efficiently analyze the collected data. It can be beneficial in tele-dermatology [18]. Here, patients need not to be present in front of any physician or expert. It produces a virtual environment and creates an illusion of the presence of the patient in the clinic. It can be helpful for those persons (especially for women) who belong to some society where exposing skin to a doctor is embarrassing or not permitted by their religion. IoT has the solution to this problem. Diagnosis is possible without the physical presence of the patient. So, there is no need to expose the region of the skin to the doctor [19]. There is no need to use very sophisticated hardware to acquire the data. The images can be acquired by using the camera of the mobile phone or so [20]. One such equipment can be observed in Fig. 1 [21]. Computer-aided diagnosis (CAD) methods should be efficient enough so that it can effectively extract information from the images and process it. CAD systems are very useful for many patients both in terms of cost and time. This work addresses some of the problems related with the IoT-based computation and proposes a new method for efficient classification of the skin diseases. A hybrid neural network model is developed with optimized feature space that makes the method suitable to be deployed in the IoT-based environment. The obtained results clearly show the efficiency of

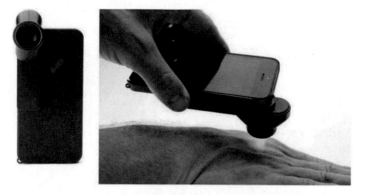

Fig. 1 Hand-held mobile dermological image acquisition device and the scanning process

the proposed method. Obtained results are so promising and therefore, the proposed method can be used in various remote healthcare systems and tele-dermatological applications. Proposed method can be used to assess the changes due to some therapy which helps in quick recovery. Developed framework enhances the classification capabilities of the IoT nodes and can be implemented with a very low cost.

3 Proposed Methodology

The proposed machine-learning-coupled IoT-based skin disease classification method is based on some distinct stages. Figure 2 illustrates the stages graphically. The functionality of the proposed frame work is divided in the following broad phases:

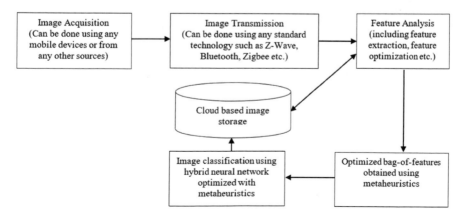

Fig. 2 System architecture

Table 1 Different image acquisition methods in dermology

Method	Pros	Cons
Digital, total body, and UV imaging	1. Simple and less costly 2. Self examination is possible	1. Only surface information reveled 2. Data is to be handled securely
Dermoscopy	1. Suitable for cancer diagnosis 2. Suitable for skin lesions diagnosis	1. Constrained magnification 2. Expertise required
RCM (reflectance confocal microscopy)	1. High resolution and depth of the acquired images 2. Suitable for histopathology 3. Biopsy can be avoided 4. Noninvasive method	1. Not reliable for deep tumors 2. Expertise required
OCT (optical coherence tomography)	1. High resolution and depth 2. Biopsy can be avoided 3. Noninvasive 4. Can produce 3D images	1. Expensive 2. Not extremely reliable for deep tumors

1. Image acquisition (i.e., Data acquisition)
2. Transmission of the acquired images
3. Analysis of the collected image and classify (includes the steps feature analysis, optimized bag of feature, and image classification using hybrid neural network. Please refer Fig. 2).

Transfer the results into some IoT-based cloud storage.

3.1 Image Acquisition

In dermology, image of the infected region is very important. In the image acquisition phase, the image of the skin with infected region is captured. Different devices can be used to capture the skin images. Due to the advancement of technology, several deices are available that can be used to acquire images with a very low cost [22, 23]. Smartphones are efficient enough so that it can be used to capture images as discussed in Sect. 2. Different techniques used for image acquisition in dermology are discussed in Table 1 [24].

3.2 Transmission of the Acquired Images

In this phase, the data is transmitted to the nearest server or data center or some organization like hospital, laboratory, etc. It is a real-time process. The transmission process uses some standard protocols like Bluetooth or Zigbee. Power is one of the major constraints in transmitting the signal [25]. Redundant data can increase the transmission overhead and can consume significant amount of power. In this work, features are selected from the captured images and reduced further to remove insignificant data. Power optimization during the transmission process is one of the active research domains. In some frameworks, concentrator modules are used to provide large-distance communication capability.

3.3 Analysis of the Collected Image and Classification

As discussed earlier, analysis and classification are the major components of IoT-based healthcare systems. It is one of the inevitable parts of the IoT-based healthcare systems because manual examination of the collected data is not always possible. Moreover, results obtained by the manual examination of the data often contaminated by some inherent errors. In the IoT environment, power utilization is one of the major issues associated with the processing nodes. Therefore, efficient and optimized classification and data interpretation methods are required to produce accurate results within stipulated time and limited computing resources [26]. In this work, a smart framework is developed that can be deployed in an IoT-based healthcare system to efficiently classify the skin lesion captured by the simple mobile image sensing devices. The detailed discussion of the proposed skin lesion analysis and classification method is given in Sect. 4.

3.4 Transfer the Results into Some IoT-Based Cloud Storage

After computation and during computation, cloud-based data storage plays a vital role. It helps to store the computed results as well as the inputted data. Physicians can access and analyze these data at any time they want. Sometimes, distributed data storages can be used to store these data [27]. The data is sometimes placed in some storage areas near to the users which are known as the cloudlets. These nodes can also perform small processing on the stored data.

4 Proposed Machine-Learning-Coupled IoT-Based Method for Automated Disease Detection from Biomedical Images

Computer-aided diagnostic methods are highly reliable and are capable to perform many intensive computations on various biomedical data [2, 28–30]. Automated disease classification is generally depending on three things. First one is the rich dataset. Dataset plays an important role in any type of classification job. Well-prepared dataset helps to obtain better convergence with obtaining higher accuracy [31]. Second important thing is the relevant features [32, 33]. Images cannot be directly analyzed by a classifier. (Note: It is possible by using deep-learning methods, e.g., convolutional neural networks which is one of the most advance topics in the domain of the computer vision. But here, our main focus is the hybrid neural networks.) Some meaningful numerical values should be extracted from the images so that classifier can work on it. Here, the features come into picture. Relevant features are very useful in classification whereas irrelevant or noisy features can degrade the overall performance. Features should be efficiently handled and reduced so that important features are preserved and redundant features are eliminated. In IoT-based framework, feature handling is an essential step because noisy features can increase the processing and transmission overhead which in turn decreases the battery life of the nodes. In this work, feature space is optimized using genetic algorithm that assures quick convergence and better performance in IoT-based infrastructure. And the third one is the quality of the classifier. It is required to design a classifier in such a way so that it can quickly update the weights and efficiently learn the model, otherwise convergence will be delayed with poor accuracy and lots of computational resources will be required. So, in this work, a hybrid artificial neural network model is proposed which can optimize the weights using some metaheuristic methods and increases the accuracy with lesser computational overhead which can be helpful in intelligent and optimized data analysis using stipulated resources. The proposed method is simple to implement and less complex than the modern machine-learning approaches.

4.1 Description of the Dataset

The dataset used in this work is collected from International Skin Imaging Collaboration (ISIC) dataset [34]. Four different types of skin diseases are considered for classification. These classes are angioma, basal cell carcinoma, lentigo simplex, and solar lentigo. The proposed model can also be used with some other classes of skin diseases. But for the simplicity, only four classes are chosen for demonstration. Images of all four types of skin diseases under consideration are given in Fig. 3. Early detection of skin cancer can cure the disease in most of the cases which reduces the mortality rate due to skin diseases. It is an open-source archive that can be used for detection and identification of the types of skin diseases from the digital images.

4.2 Feature Extraction

This step is used to extract useful and relevant features from the data, i.e., images. Images are constructed with different pixel values where the number of pixels can be

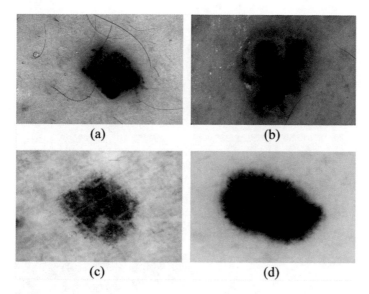

(a) (b) (c) (d)

Fig. 3 Different types of skin diseases: **a** angioma, **b** basal cell carcinoma, **c** lentigo simplex, and **d** solar lentigo

huge for many images. So, feature extraction process can reduce this huge amount of data and can map it to a smaller feature space [35]. Feature extraction should be performed in an efficient way so that the redundancy becomes minimum because redundant and irrelevant data can degrade the performance of the classifier and consumes more computational resources which can be expensive in IoT-based frameworks. Efficient feature extraction and reduction method can summarize the whole dataset and present it in front of the classifier in a comprehensive manner.

Different orientations and scales are associated with the skin images. Same images should not be classified as different due to the orientation and scale. So, to make the process reliable, the framework must be capable to detect similar images irrespective of the scale and orientation. Therefore, in this work, the scale invariant feature transform (SIFT) [36] method is used. SIFT can perform both, locate important feature points and extract features. After detecting an important point (i.e., the key point), SIFT produces a descriptor to describe that point. A typical SIFT descriptor has 128 dimensions. SIFT is proved to be very efficient and not sensitive to rotation, orientation, and scaling which is very important in skin disease analysis. A difference of Gaussian (DoG) pyramid of 'k' images is generated by the SIFT algorithm. Here, the value of the 'k' is computed from the product of the level count and octave count. It is shown in Fig. 4. It is necessary to compute the DoG pyramid to maintain the scale invariant property by determining maxima with respect to different scales. The 'k' images are obtained after computing the DoG pyramid using the image shown in Fig. 5a. SIFT first detects a huge number of key points (as shown in Fig. 5b). In this case, SIFT detects 2664 key points, i.e., the algorithm returns a matrix of dimension 2664×128 because each descriptor is of 128 dimensions. Now, this is a

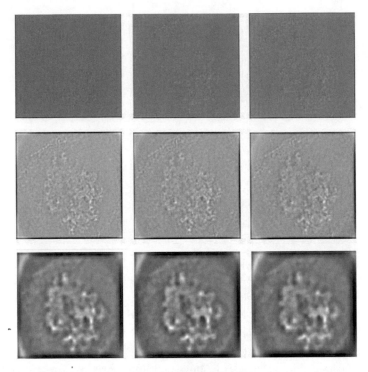

Fig. 4 DoG pyramid obtained during the feature extraction phase

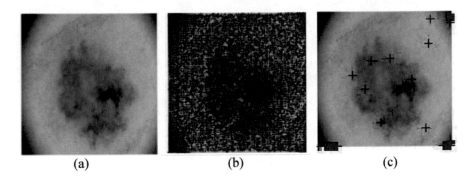

    (a)   (b)   (c)

Fig. 5 **a** Original image, **b** initial detected key points, **c** final key points

huge volume of data which is difficult to process and transmit. So, SIFT has removed the redundant keypoints and preserved the important ones as shown in Fig. 5c. For this test case, the number of features is reduced to only 53. Now, handling 53×128 dimensional matrix is comparatively computationally less expensive than 2664×128 dimensional matrix. In several cases, large amount of feature points may lead to over fitting which degrade the performance of the classifier. Therefore, feature reduction is an important step which makes the rest of the process computationally inexpensive.

4.3 Dimensionality Reduction Using Bag of Features

Although SIFT reduces the number of key points, still, the classifier needs to handle a dataset with large dimension and therefore the classifier demands high computational resources to learn the model. Moreover, it can lead to ambiguity and classifier can learn a poor model which with less generalization power [7]. Therefore, dimensionality reduction is one of the necessary steps to reduce the computational overhead in the IoT nodes that can save substantial part of the energy. In general, IoT processing nodes do not have huge computational power and resources. So, the algorithm should be smart enough so that it can reduce the computational overhead and resource utilization. In this work, 128 dimensions are reduced and mapped to a feature space of lesser dimensions [37]. In the reduced feature space, each feature is used to represent one or more than one features of the original feature space. This mapping is done by the evolutionary clustering method. Every feature represents a data point and cluster centers are updated using the genetic algorithm [38]. It is one of the simplest and elegant nature-inspired metaheuristic algorithm that can be easily implemented [39–42] and computationally inexpensive. There are various sophisticated metaheuristic algorithms, that can be found in the literature, are proven to be more powerful than genetic algorithm but are more complex and uses more computation resources. One of the major objectives of this method is to get high classification accuracy satisfying the constraints of the IoT-based computational environment. Therefore, genetic algorithm is a good tradeoff between computational overhead efficiency. Here, roulette wheel selection method [40] is used to select the individuals. Values of different parameters used for genetic algorithm is given in Table 2. The number of offsprings is computed using (1).

$$nOffsprings = 2 * round((cp * nPop)/2) \qquad (1)$$

Table 2 Values of the different parameters used for the genetic algorithm

Parameter	Values
Maximum number of iterations	200
Size of the population ($nPop$)	100
Percentage of crossover (cp)	80%
Number of offsprings ($nOffsprings$)	80
Percentage of mutation (mp)	30%
Number of mutants ($nMutants$)	30
Selection pressure	8

Here, nOffsprings is the number of off springs, cp is the percentage of the crossover, and nPop is the number of members in a population. Number of mutants can be computed using (2).

$$n\text{Mutants} = \text{round}(\text{mp} * n\text{Pop}) \tag{2}$$

Here, nMutants represents the number of mutants and mp represents the percentage of mutation.

Genetic algorithm optimizes the cost function which indicates the quality of the clustering. In this work, the Davies–Bouldin index is used as the cost function. It is given in (3).

$$\text{DBIndex} = \frac{1}{C_n} \sum_{i=1}^{C_n} P_i \tag{3}$$

Here, C_n denotes the number of clusters and P_i is represented using (4).

$$P_i = \max_{j \neq i} M_{i,j} \tag{4}$$

Here, the term $M_{i,j}$ is denoted by (5).

$$M_{i,j} = \frac{\text{IntraSep}_i + \text{IntraSep}_j}{K_{i,j}} \tag{5}$$

Here, IntraSep is defined as (6) and $K_{i,j}$ is defines as (7).

$$\text{IntraSep}_i = \left(\frac{1}{\text{SZ}_i} \sum_{j=i}^{\text{SZ}_i} |F_j - \text{Cen}_i|^p \right)^{1/p} \tag{6}$$

IntraSep actually measures the intra-cluster separation. IntraSep$_i$ denotes the separation value, SZ$_i$ denotes the size of the ith cluster, and F_j is the vector that represents a point assigned to the cluster C_i. Cen$_i$ denotes the centroid of the ith cluster C_i. If the value of the parameter 'p' is taken as 2, then the distance metric becomes Euclidean distance measure.

$$K_{i,j} = \|\text{Cen}_i - \text{Cen}_j\| = \left(\sum_{t=1}^{n} |a_{t,i} - a_{t,j}|^p \right)^{1/p} \tag{7}$$

$K_{i,j}$, denotes the inter-cluster separation and $a_{t,i}$ denotes the tth element (with n dimensions) of Cen$_i$.

The genetic algorithm tries to minimize the function given in (3). It is a minimization problem because lesser value of DBIndex indicates better inter-cluster separation, i.e., inter-cluster distance and tight bonding, i.e., lesser intra-cluster distance. The value of the parameter 'p' can be changed to implement different distance metrics. It should be decided depending on the algorithm used for clustering. Wrong choice of the distance metric can leads to poor clustering result which in turn affects the classification accuracy. This process helps in dimensionality reduction by mapping one or more than one dimensions to a single one. It will be helpful in big data handling and useful in optimizing computational power which is a major constraint in IoT-based environment [43, 44]. Genetic algorithm eliminates the need for exhaustive search and guides the optimization process that will try to get the optimum cost function. Elitism is incorporated to preserve the best solution found till time [45]. From the convergence curve, it is clear that, at any step, the best solution is preserved through elitism.

4.4 Metaheuristic-Supported Artificial Neural Network (ANN)

Learning environment is one of the most important factors that influence the accuracy of the system. The local optimization algorithms prematurely converge to local optima resulting in an ANN having unsuitable set of weights in different neural connections [46]. General gradient descent optimization methods often suffer from this issue. In this work, this issue is addressed and to overcome this problem, three metaheuristic algorithms, namely GA, flower pollination algorithm (FPA) and NSGA-II are employed to train the ANN. Conventional methods generally work on differentiable and continuous objective functions but metaheuristic approaches do not have this limitations. Metaheuristic-trained artificial neural networks can efficiently sole non-linear, non-differentiable complex problems. But no optimizer is available that can act as an universal one and can be applied to every class of problems. This fact is established in a well-known theorem called 'no free lunch theorem' [47]. The general approach to formulate metaheuristic-supported ANN is to search the optimum weights. FPA is a recently proposed metaheuristic which is inspired by the biotic and abiotic pollination of flowers. Studies have revealed its importance in training ANNs [48]. On the other hand, GA and NSGA-II are single and multi-objective global optimization techniques, respectively. The potential of NSGA-II is addressed in literature [49]. The training process of artificial neural network with the help of metaheuristic algorithm is illustrated in Fig. 6. This method can efficiently optimize the classification process and can generate better results in a stipulated time.

Fig. 6 Training method of
ANN using metaheuristic
algorithm

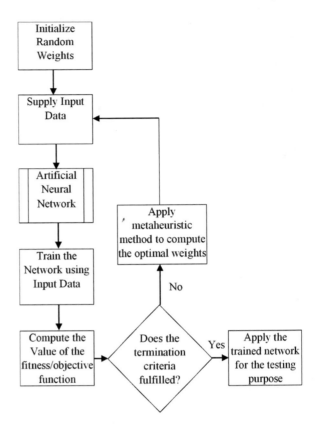

5   Results and Discussion

In the current work, feature are extracted using the SIFT method and then optimized
using the genetic algorithm. After generating the optimized features, three different
metaheuristic methods are employed to train ANNs. The experiment is conducted
using 200 samples (50 samples for each class) and the performance of the proposed
method is assessed using some standard numerical parameters. 40% of the samples
are used for the training purposes and rest of the samples is used to test the proposed
model. Some of the classes do not have 50 samples. Therefore, some samples are
repeated for those classes where sufficient amount of samples are not found. The
first ANN is trained using genetic algorithm (ANN-GA) [46], the next model of
ANN is trained with flower pollination algorithm (ANN-FPA) [50] and finally, the
ANN is trained with NSGA-II (ANN-NSGA-II) [51]. The objective functions and
algorithmic setup are as in [46, 49–51]. The same for ANN-NSGA-II is as in [49].
Tenfold cross-validation is used in experiments. Each model is run for 50 times and
the average results are reported for each model. The models are compared in terms of
following test-phase confusion-matrix-based performance metrics which are given
in (8) to (11), namely accuracy, precision, recall, and f-measure, respectively.

$$\text{Accuracy} = \frac{TP + TN}{TP + FP + TN + FN} \tag{8}$$

$$\text{Precision} = \frac{TP}{TP + FP} \tag{9}$$

$$\text{Recall} = \frac{TP}{TP + FN} \tag{10}$$

$$f\text{-Measure} = \frac{2 \times \text{Precision} \times \text{Recall}}{\text{Precision} + \text{Recall}} \tag{11}$$

The genetic algorithm reduces the dimensions by optimizing the Davies–Bouldin index. The convergence curves of the genetic algorithm for all four classes are given in Fig. 7. From the convergence curves, it is clear that the optimized Davies–Bouldin index is achieved within 200 iterations for each class and hence optimal number of dimensions can be achieved easily. These curves show the rate of convergence with respect to iterations. The best cost, i.e., minimum value of the Davies–Bouldin index is plotted in 'Y'-axis and the number of iterations is plotted in 'X'-axis. The convergence curves prove that the proposed framework overcomes any premature convergence and reach smoothly to the near global optimum and can find the reduce feature space efficiently.

Table 3 reports the performance of different metaheuristic-coupled ANN classifiers. The results are reported after performing the cross-validation step. The performance of ANN-GA is moderate with an accuracy of 72.02, precision 83.75, recall 85.9, and f-measure 84.8. ANN supported by FPA achieved better accuracy. However,

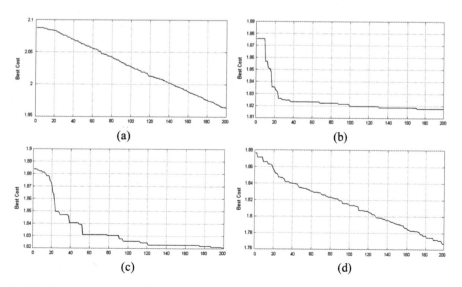

Fig. 7 Convergence curves of four different classes: **a** angioma, **b** basal cell carcinoma, **c** lentigo simplex and **d** solar lentigo

Table 3 Comparative study of metaheuristic-supported ANN classifiers

Parameters	ANN-GA	ANN-FPA	ANN-NSGA-II
Accuracy	72.02	81.26	89.47
Precision	83.75	86.5	89.19
Recall	85.9	90.91	99.08
f-measure	84.81	88.65	94.29

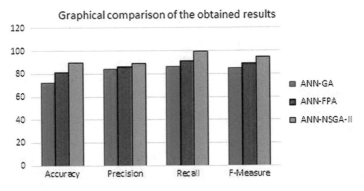

Fig. 8 Graphical comparison of the obtained results

in terms of precision, recall, and f-measure, the performance of ANN-FPA is slightly better than ANN-GA. The performance is highly improved while ANN is trained with NSGA-II. It achieves an accuracy of 89.47, precision 89.19, recall 99.08, and f-measure 94.29. The performance is significantly better than other well-known classifiers. The graphical comparison of the obtained results can be observed in Fig. 8. So, NSGA-II-coupled ANN can be used to achieve better accuracy.

Metaheuristic algorithms perform better than conventional algorithms in terms of exploitation and exploration of the search space. But, a specific method cannot be applied for every problems, hence there is a scope of improvisation to train artificial neural networks. Different metaheuristic optimization methods can be tested and applied to other modalities. From the experimental results, it can be concluded that the artificial neural network trained with NSGA-II performs better followed by flower pollination algorithm and genetic algorithm for the proposed framework where the feature space is optimized with the genetic algorithm. Other metaheuristic algorithms can be useful in some other arrangements but this framework can be directly applied on different problems besides automated skin disease identification systems.

6 Conclusion

The major goal of this work is to create an automated healthcare system that can be used in the IoT framework. The system must be capable to exploit the power of IoT within stipulated computing resources. In this work, an IoT-based framework is proposed to automatically diagnose skin diseases using computer-aided systems. Feature space is optimized using genetic algorithm to decrease the burden from the shoulder of the artificial neural network and to reduce the computational and transmission overhead from the nodes of the IoT infrastructure. It is certainly helpful in saving the power consumption. Moreover, to achieve better and faster convergence, various metaheuristic algorithms are employed to update the weights of the artificial neural networks. The result of this work is very promising which allows this algorithm to be considered for deployment in the IoT-based healthcare systems. This framework can handle big data efficiently in IoT framework. This algorithm avoids exhaustive search in both the phases, i.e., in feature space optimization and weight manipulation which save precious time and computational resources. Tenfold cross-validation makes this model reliable and suitable for real-world deployment. Still, there are some chances of improvement that can be done by extending this work. Different metaheuristic optimization algorithms can be tested on both phases to get the better performance in reasonable time. As discussed earlier, timely diagnosis is necessary for any automated diagnosis and healthcare system. Moreover, further research can be carried out to explore more on this domain that can bring a paradigm shift and can cope up with new challenges. So, this system can be practically deployed in other IoT-based environments (i.e., except dermatological image analysis) to perform automated disease detection or some other tasks where the amount of data is a nightmare. The proposed framework is a generalized one and independent of the underlying architecture, hence it can be very useful in IoT-based remote healthcare systems.

References

1. Kos, A., & Umek, A. (2018). Wearable sensor devices for prevention and rehabilitation in healthcare: Swimming exercise with real-time therapist feedback. *IEEE Internet of Things Journal,* 1. [Internet]. Cited November 8, 2018. Available from: https://ieeexplore.ieee.org/document/8395365/.
2. Chakraborty, S., & Mali, K. (2018). Application of multiobjective optimization techniques in biomedical image segmentation—A study. In *Multi-objective optimization* (pp. 181–194). Singapore: Springer Singapore. [Internet]. Cited September 14, 2018. Available from: http://link.springer.com/10.1007/978-981-13-1471-1_8.
3. Hahn, C., Kwon, H., & Hur, J. (2018). Trustworthy delegation toward securing mobile healthcare cyber-physical systems. *IEEE Internet of Things Journal,* 1. [Internet]. Cited November 8, 2018. Available from: https://ieeexplore.ieee.org/document/8510797/.
4. Pinet, É. (2009). Fabry-Pérot fiber-optic sensors for physical parameters measurement in challenging conditions. *Journal of Sensors,* 1–9. [Internet]. Cited May 11, 2019. Available from: http://www.hindawi.com/journals/js/2009/720980/.

5. Verma, N., & Singh, D. (2018). Data redundancy implications in wireless sensor networks. *Procedia Computer Science, 132*, 1210–1217. [Internet]. Cited May 6, 2019. Available from: https://www.sciencedirect.com/science/article/pii/S1877050918307683.
6. Chakraborty, S., Mali, K., Chatterjee, S., Banerjee, S., Mazumdar, K. G., Debnath, M., et al. (2017). Detection of skin disease using metaheuristic supported artificial neural networks. In *2017 8th Annual Industrial Automation and Electromechanical Engineering Conference IEMECON 2017* (pp. 224–229).
7. Chakraborty, S., Mali, K., Chatterjee, S., Anand, S., Basu, A., Banerjee, S., et al. (2017). Image based skin disease detection using hybrid neural network coupled bag-of-features. In *2017 IEEE 8th Annual Ubiquitous Computing, Electronics and Mobile Communication Conference* (pp. 242–246). IEEE. [Internet]. Cited March 13, 2018. Available from: http://ieeexplore.ieee.org/document/8249038/.
8. Greenes, R. A. (1986). Computer-aided diagnostic strategy selection. *Radiologic Clinics of North America, 24*, 105–120. [Internet]. Available from: http://ovidsp.ovid.com/ovidweb.cgi?T=JS&PAGE=reference&D=med2&NEWS=N&AN=3515404%5Cn http://ovidsp.ovid.com/ovidweb.cgi?T=JS&PAGE=reference&D=emed1b&NEWS=N&AN=1986237550.
9. Chi, Q., Yan, H., Zhang, C., Pang, Z., & Da Xu, L. (2014). A reconfigurable smart sensor interface for industrial WSN in IoT environment. *IEEE Transactions on Industrial Informatics, 10*, 1417–1425. [Internet]. Cited May 11, 2019. Available from: http://ieeexplore.ieee.org/document/6742595/.
10. Gulati, S., & Bhogal, R. K. (2020). *Classification of melanoma from dermoscopic images using machine learning* (pp. 345–354). Singapore: Springer. Cited October 12, 2019. Available from: http://link.springer.com/10.1007/978-981-13-9282-5_32.
11. Joshi, A. D., Manerkar, S. S., Nagvekar, V. U., Naik, K. P., Palekar, C. G., Pugazhenthi, P. V., et al. (2019). Skin disease detection and classification. *International Journal of Advanced Engineering Research and Science, 6*, 396–400. [Internet]. Cited October 12, 2019. Available from: https://ijaers.com/detail/skin-disease-detection-and-classification/.
12. Sarkar, R., Chatterjee, C. C., & Hazra, A. (2019). *A novel approach for automatic diagnosis of skin carcinoma from dermoscopic images using parallel deep residual networks* (pp. 83–94). Singapore: Springer. Cited October 12, 2019. Available from: http://link.springer.com/10.1007/978-981-13-9939-8_8.
13. Wang, Y., Pan, H., Yang, B., Bian, X., & Cui, Q. (2019). *Mutual learning model for skin lesion classification* (pp. 214–222). Singapore: Springer. Cited October 12, 2019. Available from: http://link.springer.com/10.1007/978-981-15-0121-0_17.
14. Andersen, C. (2019). *Melanoma classification in low resolution dermoscopy images using deep learning*. University of Bergen. [Internet]. Cited October 12, 2019. Available from: http://bora.uib.no/bitstream/handle/1956/20304/Master-Thesis-By-Christian_Andersen.pdf?sequence=1.
15. Gu, Y., Ge, Z., Bonnington, C. P., & Zhou, J. (2019). Progressive transfer learning and adversarial domain adaptation for cross-domain skin disease classification. *IEEE Journal of Biomedical and Health Informatics, 1*. [Internet]. Cited October 12, 2019. Available from: https://ieeexplore.ieee.org/document/8846038/.
16. Chatterjee, S., Dey, D., Munshi, S., & Gorai, S. (2019). Extraction of features from cross correlation in space and frequency domains for classification of skin lesions. *Biomedical Signal Processing and Control, 53*, 101581. [Internet]. Cited October 12, 2019. Available from: https://www.sciencedirect.com/science/article/abs/pii/S1746809419301557.
17. Jebadurai, J., & Dinesh Peter, J. (2018). Super-resolution of retinal images using multi-kernel SVR for IoT healthcare applications. *Future Generation Computer Systems, 83*, 338–346. [Internet]. Cited June 10, 2019. Available from: https://www.sciencedirect.com/science/article/pii/S0167739X17322136.
18. *What is teledermatology? Virtual dermatology practice*. eVisit. [Internet]. Cited November 1, 2018. Available from: https://evisit.com/resources/what-is-teledermatology/.
19. Desai, B., McKoy, K., & Kovarik, C. (2010). Overview of international teledermatology. *Pan African Medical Journal, 6*, 3. [Internet]. Cited November 1, 2018. Available from: http://www.ncbi.nlm.nih.gov/pubmed/21436946.

20. Bourouis, A., Zerdazi, A., Feham, M., & Bouchachia, A. (2013). M-health: Skin disease analysis system camera. *Procedia Computer Science, 19*, 1116–1120. [Internet]. Cited November 1, 2018. Available from: www.sciencedirect.com.

21. *Handheld dermoscopy for skin cancer detection*. [Internet]. Cited November 1, 2018. Available from: https://dermlite.com/.

22. Chakraborty, S., Chatterjee, S., Dey, N., Ashour, A. S., Ashour, A. S., Shi, F., et al. (2017). Modified cuckoo search algorithm in microscopic image segmentation of hippocampus. *Microscopy Research and Technique, 80*, 1051–1072. [Internet]. Cited September 14, 2018. Available from: http://doi.wiley.com/10.1002/jemt.22900.

23. Hore, S., Chakroborty, S., Ashour, A. S., Dey, N., Ashour, A. S., Sifaki-Pistolla, D., et al. (2015). Finding contours of hippocampus brain cell using microscopic image analysis. *Journal of Advanced Microscopy Research, 10*, 93–103. [Internet]. Cited January 14, 2018. Available from: http://openurl.ingenta.com/content/xref?genre=article&issn=2156-7573&volume=10&issue=2&spage=93.

24. Hibler, B., Qi, Q., & Rossi, A. (2016). Current state of imaging in dermatology. *Seminars in Cutaneous Medicine and Surgery, 35*, 2–8. [Internet]. Cited November 1, 2018. Available from: http://scmsjournal.com/article/abstract/current-state-of-imaging-in-dermatology/.

25. Luo, J., Hu, J., Wu, D., & Li, R. (2015). Opportunistic routing algorithm for relay node selection in wireless sensor networks. *IEEE Transactions on Industrial Informatics, 11*, 112–121. [Internet]. Cited May 11, 2019. Available from: http://ieeexplore.ieee.org/document/6965597/.

26. Chakraborty, S., Roy, M., & Hore, S. (2018). A study on different edge detection techniques in digital image processing. In *Computer vision: Concepts, methodologies, tools, and applications*.

27. Verma, P., & Sood, S. K. (2018). Fog assisted-IoT enabled patient health monitoring in smart homes. *IEEE Internet of Things Journal, 5*, 1789–1796. [Internet]. Cited November 8, 2018. Available from: https://ieeexplore.ieee.org/document/8283747/.

28. Chakraborty, S., Chatterjee, S., Ashour, A. S., Mali, K., & Dey, N. (2017). Intelligent computing in medical imaging: A study. In N. Dey (Ed.), *Advancements in applied metaheuristic computing* (pp. 143–163). IGI Global. [Internet]. Cited November 24, 2017. Available from: http://services.igi-global.com/resolvedoi/resolve.aspx?doi=10.4018/978-1-5225-4151-6.ch006.

29. Chakraborty, S., Chatterjee, S., Das, A., & Mali, K. (2020). Penalized fuzzy C-means enabled hybrid region growing in segmenting medical images (pp. 41–65).

30. Hore, S., Chakraborty, S., Chatterjee, S., Dey, N., Ashour, A. S., Van Chung, L., et al. (2016). An integrated interactive technique for image segmentation using stack based seeded region growing and thresholding. *International Journal of Electrical & Computer Engineering, 6*.

31. Azar, J., Makhoul, A., Barhamgi, M., & Couturier, R. (2019). An energy efficient IoT data compression approach for edge machine learning. *Future Generation Computer Systems, 96*, 168–175. [Internet]. Cited June 10, 2019. Available from: https://www.sciencedirect.com/science/article/pii/S0167739X18331716.

32. Chakraborty, S., Mali, K., Chatterjee, S., Anand, S., Basu, A., Banerjee, S., et al. (2017). Image based skin disease detection using hybrid neural network coupled bag-of-features. In *2017 IEEE 8th Annual Ubiquitous Computing, Electronics and Mobile Communication Conference* (pp. 242–246). IEEE. [Internet]. Cited March 19, 2018. Available from: http://ieeexplore.ieee.org/document/8249038/.

33. Chakraborty, S., Chatterjee, S., Chatterjee, A., Mali, K., Goswami, S., & Sen, S. (2018). Automated breast cancer identification by analyzing histology slides using metaheuristic supported supervised classification coupled with bag-of-features. In *2018 Fourth International Conference on Research in Computational Intelligence and Communication Networks* (pp. 81–86). IEEE. [Internet]. Cited June 10, 2019. Available from: https://ieeexplore.ieee.org/document/8718736/.

34. ISIC Archive. [Internet]. Cited November 1, 2018. Available from: https://www.isic-archive.com/#!/topWithHeader/tightContentTop/about/isicArchive.

35. Hore, S., Chatterjee, S., Chakraborty, S., & Kumar Shaw, R. (2016). Analysis of different feature description algorithm in object recognition. In *Feature detectors and motion detection in video*

processing (pp. 66–99). IGI Global. [Internet]. Available from: https://www.igi-global.com/chapter/analysis-of-different-feature-description-algorithm-in-object-recognition/170213.

36. Lowe, D. G. (2004). Distinctive image features from scale-invariant keypoints. *International Journal of Computer Vision, 60*, 91–110. [Internet]. Cited November 1, 2018. Available from: http://link.springer.com/10.1023/B:VISI.0000029664.99615.94.

37. Chakraborty, S., Mali, K., Banerjee, S., Roy, K., Saha, D., Chatterjee, S., et al. (2017). Bag-of-features based classification of dermoscopic images. In *2017 4th International Conference on Opto-Electronics and Applied Optics* (pp. 1–6). IEEE. [Internet]. Cited July 16, 2018. Available from: https://ieeexplore.ieee.org/document/8349977/.

38. Holland, J. H. (1992). Genetic algorithms. *Scientific American, 267*, 66–72. [Internet]. Available from: http://www.nature.com/doifinder/10.1038/scientificamerican0792-66.

39. Chakraborty, S., & Bhowmik, S. (2013). Job shop scheduling using simulated annealing. In *First International Conference on Computation and Communication Advancement* (pp. 69–73). McGraw Hill Publication. [Internet]. Cited November 24, 2017. Available from: https://scholar.google.co.in/citations?user=8lhQFaYAAAAJ&hl=en.

40. Chakraborty, S., & Bhowmik, S. (2015). Blending roulette wheel selection with simulated annealing for job shop scheduling problem. In *Michael Faraday IET International Summit 2015* (p. 100 (7.)). Institution of Engineering and Technology. [Internet]. Cited November 24, 2017. Available from: http://digital-library.theiet.org/content/conferences/10.1049/cp.2015.1696.

41. Chakraborty, S., & Bhowmik, S. (2015). An efficient approach to job shop scheduling problem using simulated annealing. *International Journal of Hybrid Information Technology, 8*, 273–284. [Internet]. Cited November 24, 2017. Available from: http://dx.doi.org/10.14257/ijhit.2015.8.11.23.

42. Chakraborty, S., Seal, A., & Roy, M. (2015). An elitist model for obtaining alignment of multiple sequences using genetic algorithm. *International Journal of Innovative Research in Science, Engineering and Technology, 61*–67.

43. Magno, M., Boyle, D., Brunelli, D., Popovici, E., & Benini, L. (2014). Ensuring survivability of resource-intensive sensor networks through ultra-low power overlays. *IEEE Transactions on Industrial Informatics, 10*, 946–956. [Internet]. Cited May 11, 2019. Available from: http://ieeexplore.ieee.org/document/6687258/.

44. Rani, S., Ahmed, S. H., Talwar, R., & Malhotra, J. (2017). Can sensors collect big data? An energy-efficient big data gathering algorithm for a WSN. *IEEE Transactions on Industrial Informatics, 13*, 1961–1968. [Internet]. Cited May 11, 2019. Available from: http://ieeexplore.ieee.org/document/7829299/.

45. Chakraborty, S., Raman, A., Sen, S., Mali, K., Chatterjee, S., & Hachimi, H. (2019). Contrast optimization using elitist metaheuristic optimization and gradient approximation for biomedical image enhancement. In *2019 Amity International Conference on Artificial Intelligence* (pp. 712–717). IEEE. [Internet]. Cited May 12, 2019. Available from: https://ieeexplore.ieee.org/document/8701367/.

46. Chatterjee, S., Ghosh, S., Dawn, S., Hore, S., & Dey, N. (2016). Forest type classification: A hybrid NN-GA model based approach. In *Advances in Intelligent Systems and Computing*.

47. Wolpert, D. H., & Macready, W. G. (1997). No free lunch theorems for optimization. *IEEE Transactions on Evolutionary Computation, 1*, 67–82. [Internet]. Cited June 10, 2019. Available from: http://ieeexplore.ieee.org/document/585893/.

48. Chiroma, H., Khan, A., Abubakar, A. I., Saadi, Y., Hamza, M. F., Shuib, L., et al. (2016). A new approach for forecasting OPEC petroleum consumption based on neural network train by using flower pollination algorithm. *Applied Soft Computing, 48*, 50–58. [Internet]. Cited November 7, 2018. Available from: https://www.sciencedirect.com/science/article/pii/S1568494616303180.

49. Chatterjee, S., Sarkar, S., Dey, N., & Sen, S. (2018). Non-dominated sorting genetic algorithm-II-induced neural-supported prediction of water quality with stability analysis. *Journal of Information & Knowledge Management, 17*, 1850016. [Internet]. Cited November 7, 2018. Available from: https://www.worldscientific.com/doi/abs/10.1142/S0219649218500168.

50. Yang, X.-S., Karamanoglu, M., & He, X. (2014). Flower pollination algorithm: A novel approach for multiobjective optimization. *Engineering Optimization, 46*, 1222–1237. [Internet]. Cited November 7, 2018. Available from: http://www.tandfonline.com/doi/abs/10.1080/0305215X.2013.832237.
51. Chatterjee, S., Sarkar, S., Hore, S., Dey, N., Ashour, A. S., Shi, F., et al. (2017). Structural failure classification for reinforced concrete buildings using trained neural network based multi-objective genetic algorithm. *Structural Engineering and Mechanics, 63*, 429–438.

Transfer Learning Coupled Convolution Neural Networks in Detecting Retinal Diseases Using OCT Images

Kyamelia Roy, Sheli Sinha Chaudhuri, Probhakar Roy, Sankhadeep Chatterjee, and Soumen Banerjee

Abstract Optical coherence tomography (OCT) in diagnosing retinal images is an extensive technique for detecting the wide-ranging diseases related to retina. In this paper, the authors have considered three diseases, viz. diabetic macular edema (DME), choroidal neovascularization (CNV), and drusen. These diseases are classified using six different convolutional neural network (CNN) architectures. The purpose is to compare among the six different CNNs in terms of accuracy, precision, *F*- measure, and recall. The architectures used are coupled with or without transfer learning, and a comparison has been drawn as to how the CNN architectures work when they are coupled with or without transfer learning. A dataset has been considered with the mentioned retinal diseases and no pathology. The designed models could identify the specific disease or no pathology when fed with multiple retinal images of various diseases. The training accuracies obtained for the six architectures, viz. four convolutional layer deep CNNs, VGG (VGG-16 and VGG-19) and Google's Inception [Google's Inception v3 (with or without transfer learning)], and Google's Inception v4, are, respectively, 87.15%, 91.40%, 93.32%, 85.31%, and 83.63%, respectively, while the corresponding validation accuracies are 73.68%, 88.39%, 86.95%, 85.30%, and 79.50%. Thus, the results so obtained are promising in nature and establish the superiority of the proposed model.

K. Roy · S. S. Chaudhuri
Department of ETCE, Jadavpur University, Kolkata 700032, India
e-mail: kyamelia2015@gmail.com

S. S. Chaudhuri
e-mail: shelisinhachaudhuri@gmail.com

P. Roy · S. Banerjee (✉)
Department of ECE, University of Engineering and Management, Kolkata 700160, India
e-mail: prof.sbanerjee@gmail.com

P. Roy
e-mail: probhakarroy3110@gmail.com

S. Chatterjee
Department of Computer Science and Engineering, University of Engineering and Management, Kolkata 700160, India
e-mail: chatterjeesankhadeep.cu@gmail.com

© Springer Nature Singapore Pte Ltd. 2020
J. K. Mandal and S. Banerjee (eds.), *Intelligent Computing: Image Processing Based Applications*, Advances in Intelligent Systems and Computing 1157,
https://doi.org/10.1007/978-981-15-4288-6_10

Keywords Convolutional neural network · Choroidal neovascularization · Deep learning · Diabetic macular edema · Drusen · Optical coherence tomography (OCT) · Ophthalmology

1 Introduction

Eye diseases leading to blindness are a social menace which needs to be eradicated completely for the benefit of mankind and society at large. In developing countries, a large section of the society suffers from various eye diseases which go undiagnosed at times. The socioeconomic condition is also a burden which often prevents the patients to treat such diseases at an earlier stage. Research is being carried out worldwide to combat the visual impairment and to provide proper scientific diagnosis and subsequent treatment of the diseases related to eyes. In this context, optical coherence tomography (OCT), a noninvasive optical medical diagnostic imaging modality, has played a key role in being an integral imaging instrument in ophthalmology [1, 2]. It generates 3D or cross-sectional images through the measurement of echo time delay and magnitude of backscattered or back-reflected light. The rapid development and tremendous impact of OCT imaging in clinical diagnosis are increasing day-by-day with its first study on human retina in [3, 4]. It provides in vivo cross-sectional imaging of microstructure in biological system [5–7] and facilitates imaging of retinal structure which cannot be obtained through other noninvasive diagnostic techniques. The ophthalmic treatment proves to be one of the most clinically developed applications of OCT imaging [8, 9]. Its popularity across the globe is accounted for the availability of fourth-generation instruments and half-dozen companies commercializing this technology worldwide for ophthalmic diagnosis. Its advantages for earlier diagnosis of pathologies are accounted for its textual and morphological variations in properties [10, 11]. Thus considering the advanced features of OCT images, the authors in the present paper have considered such images of retina for eye-related diseases for analysis, detection, and subsequent classification using various architectures of convolutional neural network (CNN).

The diseases of eye considered here are diabetic macular edema (DME), choroidal neovascularization (CNV), and drusen. The aforesaid diseases have been chosen owing to their threat in causing irreversible vision loss leading to blindness in developed and developing countries [12–14]. Early detection of such diseases would definitely lead to better control and overcoming the curse of blindness. Several researchers across the globe are working on OCT-based image classification of DME [15–17], age-related macular degeneration (AMD) [18], CNV [19], and drusen [20, 21] using deep learning and other CNN tools [22]. In [23], a computer-aided diagnosis model was proposed to distinguish DME, age-related macular degeneration (AMD), and healthy macula based on linear configuration pattern (LCP) features of OCT images and correlation-based feature subset (CFS) selection algorithm. An automated system using color retinal fundus images-based feature learning approach was developed [24], for earlier diagnosis and treatment of DME. Machine learning algorithm using receiver operator characteristic (ROC) analysis and Cohen's statistics was proposed [25] for automatically grading AMD severity stages from OCT scans. An automated algorithm was proposed for CNV area detection in participants with AMD [26] while automated segmentation of CNV in OCT images was carried out for the treatment of CNV diseases [27]. The U-Net CNN architecture is applied for automated segmentation of drusen from fundus image and further the classification of early or advanced stages of AMD [28].

Artificial neural network has a plethora of applications in computer vision, speech processing, medical analysis, etc. The deep learning architectures or models are mostly supported by ANN. These architectures are implemented in various fields and have given promising results which are superior to human analysis in comparison. A deep learning method was proposed to distinguish between normal OCT retinal images and AMD affected images [29]. Likewise, an automated segmentation technique for the detection of intra-retinal fluid using deep learning method in macular OCT scans was also proposed [30]. The application of deep learning for retinal disease diagnosis in the field of ophthalmology has led to the development of a fully automated system for detection and quantification of macular fluid [31]. The results so obtained are highly perfect in terms of accuracy and precision. A popular approach in deep learning is transfer learning which has paved the path of reusing a pre-trained model on a new problem. Transfer learning is widely used in biomedical image interpretation medical decision making for eye-related diseases using retinal OCT images [32]. The retinal OCT images are classified for DME, dry AMD or no pathology based on transfer learning with pre-trained CNN GoogLeNet [33]. In the present work, the authors have used four pre-trained models, viz. Google's Inception v3, Google's Inception v4, VGG 16, and VGG 19, followed by a shallow ConvNet for the classification of the eye diseases. The results obtained are found to be very satisfactory.

2 Theoretical Background

2.1 Convolutional Neural Network

A convolutional neural network (CNN or ConvNet) is a type of deep feed-forward neural network finding a plethora of applications related to image analysis. Its receptive field property makes it more suitable for applications related to image processing in areas of biomedical imaging, remote sensing and GIS, cognitive science, etc. The partial extensions of the receptive fields of different neurons aid to cover the entire image for its successful processing. The advantage of CNN is the time complexity which is less in comparison with other image classification methods. Among several properties, an important defining property of CNN makes it shift-invariant artificial neural network (SIANN) having shared weight architecture and translation invariance

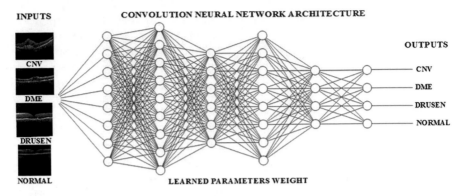

Fig. 1 Schematic diagram of CNN architecture

characteristics. In deep learning models, instead of only one complex transformation $f(D)$ (where D represents raw data), multiple transformations are adopted in sequences $(f(D) = k(g(h(D))))$ with decision boundaries present in the final layer of the CNN architecture. CNNs are universally applied in image processing, classification applications, and computer vision, wherein transformation is achieved through filters at each layer. Figure 1 depicts the schematic diagram of CNN architecture with several eye diseases as inputs and their respective classified outputs.

2.2 Motivations to Use Different CNN Architectures

The different CNN architecture used in the present work are four convolutional layer deep CNNs, VGG (VGG-16 and VGG-19), and Google's Inception (Google's Inception v3 and Google's Inception v4). They were chosen due to their increasing architectural sophistication, low error margin on the ImageNet [34] and better efficiency. Table 1 depicts the percentage of error in connection with the popularly available architectures of CNN.

Two of the above-mentioned architectures used in the work, viz. four convolutional layer deep CNNs and Google's Inception v3, are implemented with normal training

Table 1 Percentage of error in different architectures of CNN

Networks	Error (%)
AlexNet	16.0
VGG-16	7.4
VGG-19	7.3
GoogLeNet	6.7
Google's Inception v3	5.6
Google's Inception v4	5.0

and four architectures, viz. Google's Inception v3 with transfer training, Google's Inception v4 with transfer training, VGG-16 and VGG-19 with transfer training are all implemented with transfer learning. It has been observed that the accuracy in case of transfer learning is more than that for normal training.

Transfer learning method utilizes a pre-trained neural network, and to perform the operation, the final classification layer of the network is removed; the next-to-last layer of the CNN is extracted. The Inception or VGG models pre-trained with ImageNet [34] dataset to classify 1000 classes is used in the present paper as base models for the architectures using transfer learning. A shallow convolutional neural network is made as the top layer to the base model without freezing the base model. The shallow convolutional neural network comprises of the following layers:

- Batch normalization layer to increase speed and accuracy.
- ReLu activation layer.
- Dropout layer to prevent overfitting.
- Dense layer with four units/nodes (corresponding to four output classes) with softmax activation.
- Adam optimizer is used with learning rate of 0.001.

The image data is then fed from the previously built HDF5 dataset files to the base models without freezing the base models so that while training the transfer learning models, the pre-trained base models can have the ability to reconfigure some of its weights according to a new type of data fed to it for better classification purposes. This feature makes the model used in the paperless computationally expensive than Inception v3 transfer learning model used in [32]. The pre-trained convolutional base model just acts as a feature extractor in the training process without adjusting the pre-trained weights any further for different types of dataset that has quite different types of features than the ImageNet dataset, fed to it for the classification purposes. Hence with only 10 epochs of run for the used model in the paper and the usage of only 16,256 images, an accuracy of about 88% on the validation dataset is achieved. Table 2 shows the summarization of the different layers, and each of the architectures is depicted.

2.3 Training of the CNN Architectures

CNN has two cycles—forward and backward cycles. Forward cycle involves the classification of the images and computing loss (i.e., the error between the predicted label and actual label) while backward cycle involves the correction of the learnable parameters and weights based on the computed loss through a back-propagation algorithm. Backward cycle is only employed during the training phase. In the training phase, we feed the convolutional neural network architecture by batches of images with their actual labels and train them over 10 epochs.

Table 2 Number of layers in different architectures of CNN

Architecture	Layers						
	Fully connected	2D conv.	2D max pooling	2D avg. pooling	Batch norm.	Drop out	Merge
Four convolutional layer deep CNNs	4	4	4	0	0	1	0
Inception v3 with normal training	2	94	4	9	94	1	10
Inception v3 with transfer training	3	95	4	9	95	1	10
Inception v4 with transfer training	3	150	4	14	150	1	25
VGG-16 with transfer training	3	14	5	0	1	1	0
VGG-19 with transfer training	3	17	5	0	1	1	0

2.4 Dataset

The OCT image dataset [32] is used here for classification. The dataset contains about 207,130 OCT images in total out of which 108,312 OCT images are of 4686 patients (37,206 with choroidal neovascularization, 11,349 with diabetic macular edema, 8617 with drusen and 51,140 normal) with three different eye diseases. The authors have taken 16,256 OCT images (4064 with choroidal neovascularization, 4064 with diabetic macular edema, 4064 with drusen, and 4064 normal) in 127 batches to train our convolutional neural network architectures.

2.5 Python Libraries Used

TensorFlow, TensorBoard, Keras, Keras-vis, TFLearn, NumPy, H5py, OpenCV, Matplotlib, Requests, Scikit-learn, Pillow.

3 Proposed Methodology

The methodology of the present work is represented in a workflow diagram shown in Fig. 2. The workflow diagram comprises different steps involved in the present work. At the onset, the raw OCT images of the different eye diseases, shown in Fig. 3, are taken as inputs and are preprocessed by means of normalization resulting in images as shown in Fig. 4. The training and validation datasets comprising of 16,256 OCT images for training and 1000 OCT images for validation are created in the HDF5 file. The datasets are then converted into feature vectors and fed to the neural network architectures for continuous training as well as monitoring the accuracy. The training module is followed by testing where validation accuracy is measured against the unknown image data, and classifications of the diseased and

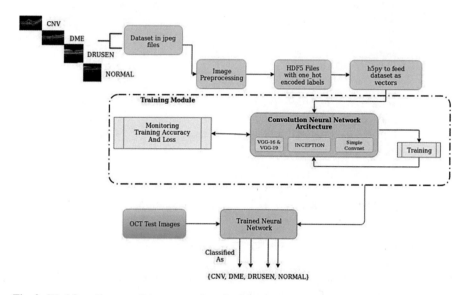

Fig. 2 Workflow diagram of the proposed methodology

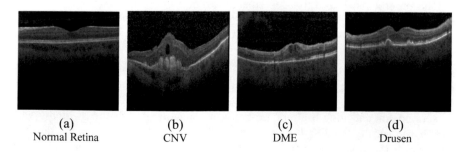

(a)	(b)	(c)	(d)
Normal Retina	CNV	DME	Drusen

Fig. 3 Input OCT images of the different eye diseases along with that of a normal retina

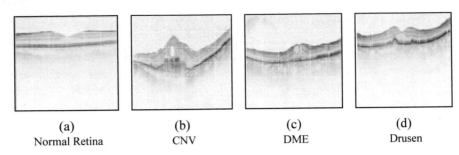

| (a) | (b) | (c) | (d) |
| Normal Retina | CNV | DME | Drusen |

Fig. 4 Preprocessed images of different eye diseases along with that of a normal retina

normal retina are achieved as the output. A brief description of the different modules of the proposed methodology is discussed below.

3.1 Image Preprocessing

The OCT images in JPEG file format are all resized into images of size 299 pixel × 299 pixel and normalized, i.e., the pixel data is mathematically divided by 255 before feeding them into a HDF5 file to be further fed into the neural network. The preprocessed images are shown in Fig. 4.

3.2 HDF5 Files with One_Hot Encoded Labels

An HDF5 file is a container for two kinds of objects: datasets, which are array-like collections of data, and groups, which are folder-like containers that hold datasets and other groups to address current and anticipated requirements of modern systems and applications. The HDF5 files help to mathematically operate on huge datasets of terabytes of data and load them into the RAM possible.

The preprocessed OCT images created are converted into two HDF5 files as training and validation datasets. The 16,256 image data forms our feature, i.e., 'X' and the type of images, i.e., CNV, DME, drusen, or normal with one_hot encoding forms our labels, i.e., 'y' stored in one HDF5 file for the training dataset. Similarly, the 1000 image data forms our feature, i.e., 'X' and the type of images, i.e., CNV, DME, drusen, or normal with one_hot encoding forms our labels, i.e., 'y' stored in one HDF5 file for the validation dataset.

An one_hot encoding is a representation of categorical variables as binary vectors. It is basically labeling the image data which defines the class of the data for perfect classification at the output stage. This first requires that the categorical values be mapped to integer values and then each integer value be represented as binary vector, of the size of the different types of categorical values (for our case, it is four representing three different classes of eye disease and one for the normal eye OCT). The binary vector representation corresponding to a type image representing diseased or normal eye is marked as binary '1' with all other values marked as binary '0.'

3.3 H5PY to Feed Dataset as Vectors

The HDF5 files created in the previous step needs the h5py module for reading purposes in a Python script. The h5py module is used to read the HDF5 dataset files and stores the image data as feature vector 'X' and one_hot encoded labels as 'y.' These features and labels are further fed into the convolutional neural network architectures in batches of 128 images for 10 epochs.

3.4 Training Module

Training accuracy gives the percentage of images being used in that training batch labeled with the correct class. Validation accuracy gives the true measure of performance of the architecture on a particular dataset which is not used in the training dataset. The total numbers of images used for training are 16,256 in 127 batches. Each batch contains 128 OCT images. The training time is approximately 1 h for the simple ConvNet with four convolutional layer deep CNNs with normal training, 4 h for Google's Inception v3 CNN with normal training and 4 h each for Google's Inception v3, Google's Inception v4, VGG-16 and VGG-19 CNN with transfer learning training. With the process in progress, accuracy involved in the training is studied.

3.5 Monitoring Training Accuracy and Loss

The training accuracy and loss of the convolutional neural network architectures can be monitored using the event log files generated by the TensorFlow framework while training the architecture and reading them with TensorBoard where all accuracy and loss metrics get plotted instantaneously. The instantaneous training accuracy and loss value can also be observed from the Keras or TFLearn metrics while training.

3.6 Trained Module

A trained neural network is obtained after the training module is continuously trained with the said architectures and performance is continuously monitored from Keras and TFLearn metrics. At the trained module, final test accuracy is performed with random set of test images. This testing gives the validation accuracy of the network and by means of which the estimation of the performance on the classification task is also evaluated. The trained module gives the classified output as CNV, DME, drusen, and normal retina which is the prime goal of the research work.

4   Results and Discussions

The different retinal diseases or malfunctions are evaluated through a comprehensive simulation study, and the proposed AI model is validated against different parameters for the different CNN architectures. All neural networks were trained on Google Colab's GPU backend with Tesla K80 GPU.

The different CNN architectures under the current study are first compared in terms of training accuracy and loss, obtained through simulation using training datasets as inputs, which are given in Table 3. The respective graphs depicting the training accuracy and loss are shown in Figs. 5, 6, 7, 8, 9, and 10. The architectures are then validated using the validation datasets, and the accuracies so obtained are given in Table 4. Thus, the already trained architectures are now capable enough to classify any given OCT retinal images into the four categories of diseases under

Table 3 Accuracy percentages and losses for the six different architectures

Models	Training metrics	
	Accuracy (%)	Loss
Four convolutional layer deep CNNs	87.15	0.42
Inception v3 with normal training	44.53	1.20
Inception v3 with transfer training	91.40	0.31
Inception v4 with transfer training	93.32	0.25
VGG-16 with transfer training	85.31	0.44
VGG-19 with transfer training	83.63	0.52

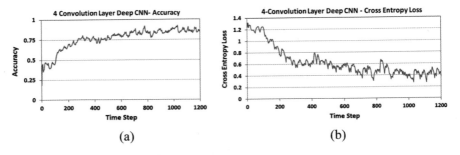

(a)   (b)

Fig. 5   Training accuracy and loss for four convolutional layer deep CNNs

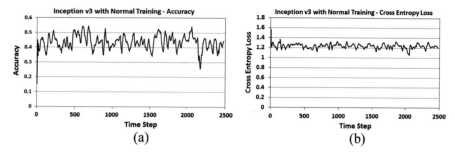

Fig. 6 Training accuracy and loss for Inception v3 with normal training

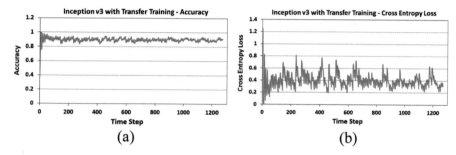

Fig. 7 Training accuracy and loss for Inception v3 with transfer learning

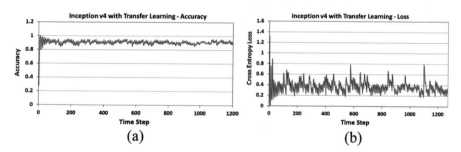

Fig. 8 Training accuracy and loss for Inception v4 with transfer learning

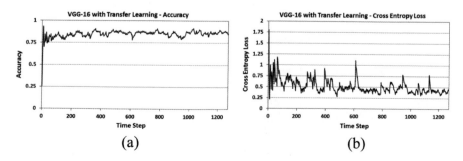

Fig. 9 Training accuracy and loss for VGG 16 with transfer learning

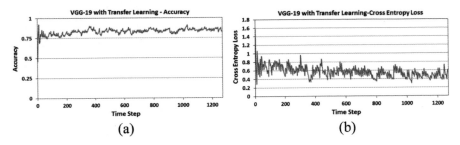

Fig. 10** Training accuracy and loss for VGG 19 with transfer learning

Table 4 Performance analysis in terms of validation accuracy

Models	Accuracy (%)
Four convolutional layer deep CNNs	73.68
Inception v3 with normal training	25.00
Inception v3 with transfer training	88.39
Inception v4 with transfer training	86.95
VGG-16 with transfer training	85.30
VGG-19 with transfer training	79.50

consideration. Now OCT images are fed to the different architectures, and the corresponding classified outputs are viewed through Python output window as shown in Figs. 9, 10, 11, 12, 13, and 14.

For the first architecture, four convolutional layer deep CNNs with normal training, the training accuracy obtained is 87.15% with 0.42 cross-entropy loss (shown in Fig. 5) as compared to the validation accuracy of 73.68%. From Fig. 11a, it is revealed that with this architecture proper classification is obtained for the diseased image 'drusen,' thereby authenticating the validation accuracy.

For the second architecture which is Google's Inception v3 CNN, the training accuracy obtained is 44.53% (shown in Fig. 6a) against a validation accuracy of 25% with normal training. The cross-entropy loss in this case is 1.2 (shown in Fig. 6b) which is very high in comparison with losses associated with other architectures. Thus owing to higher loss and lower values of training and validation accuracy, this architecture is not a promising one and leads to misclassification of images which is quite evident in case of a typical 'drusen' image being classified as 'CNV' as revealed in the Python output window as shown in Fig. 11b. However, the same architecture trained with transfer learning yields better results in terms of both training and validation accuracies of 91.4% and 88.4%, respectively, with only 0.31 cross-entropy loss (shown in Fig. 7). Hence, the classified output as shown in Fig. 11c perfectly classifies the OCT images as that of DME which authenticates our results.

Similar results are also obtained with Google's Inception v4 architecture where the training accuracy, loss, and validation accuracy are 93.32%, 0.25%, and 86.95%, respectively. The plot of training accuracy and loss is shown in Fig. 8 while the

Fig. 11 Classified results
with six architectures with
and without transfer learning

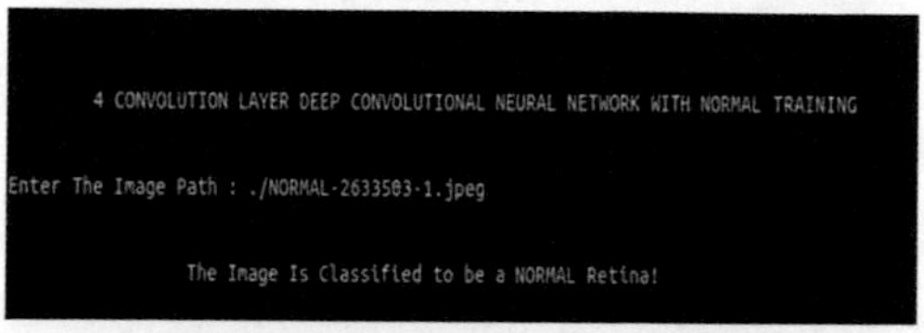

4 CONVOLUTION LAYER DEEP CONVOLUTIONAL NEURAL NETWORK WITH NORMAL TRAINING

Enter The Image Path : ./NORMAL-2633503-1.jpeg

The Image Is Classified to be a NORMAL Retina!

(a). 4-Convolution Layer Deep CNN

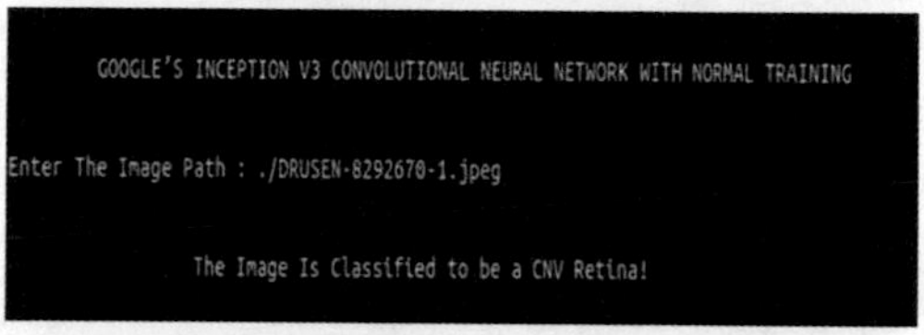

GOOGLE'S INCEPTION V3 CONVOLUTIONAL NEURAL NETWORK WITH NORMAL TRAINING

Enter The Image Path : ./DRUSEN-8292670-1.jpeg

The Image Is Classified to be a CNV Retina!

(b). Inception v3 with Normal Training

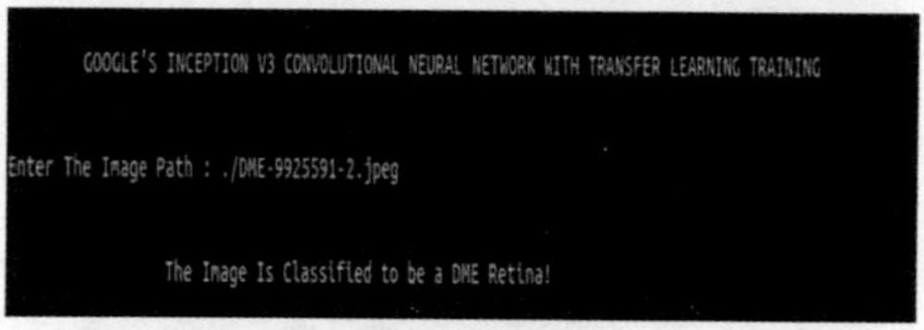

GOOGLE'S INCEPTION V3 CONVOLUTIONAL NEURAL NETWORK WITH TRANSFER LEARNING TRAINING

Enter The Image Path : ./DME-9925591-2.jpeg

The Image Is Classified to be a DME Retina!

(c). Inception v3 with Transfer Training

GOOGLE'S INCEPTION V4 CONVOLUTIONAL NEURAL NETWORK WITH TRANSFER LEARNING TRAINING

Enter The Image Path : ./CNV-8598714-1.jpeg

The Image Is Classified to be a CNV Retina!

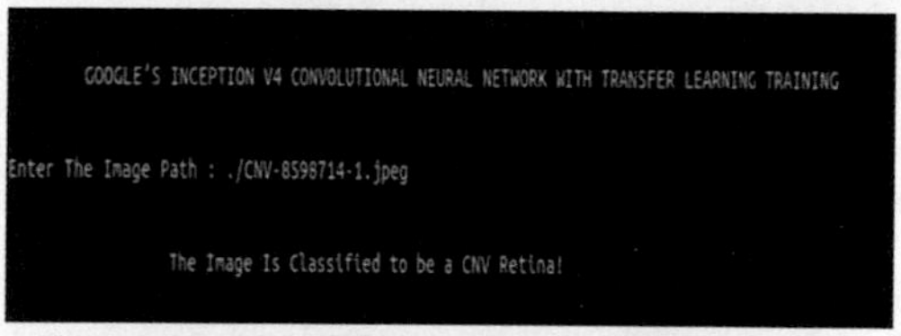

(d). Inception v4 with Transfer Training

VGG-16 CONVOLUTIONAL NEURAL NETWORK WITH TRANSFER LEARNING TRAINING

Enter The Image Path : ./CNV-28682-9.jpeg

The Image Is Classified to be a CNV Retina!

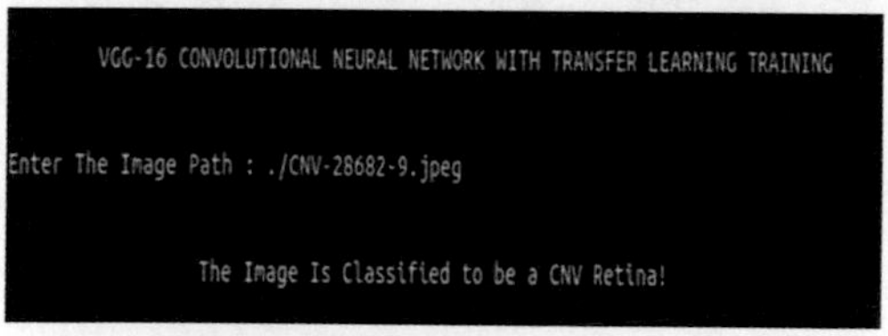

(e). VGG-16 with Transfer Training

VGG-19 CONVOLUTIONAL NEURAL NETWORK WITH TRANSFER LEARNING TRAINING

Enter The Image Path : ./CNV-8598714-1.jpeg

The Image Is Classified to be a CNV Retina!

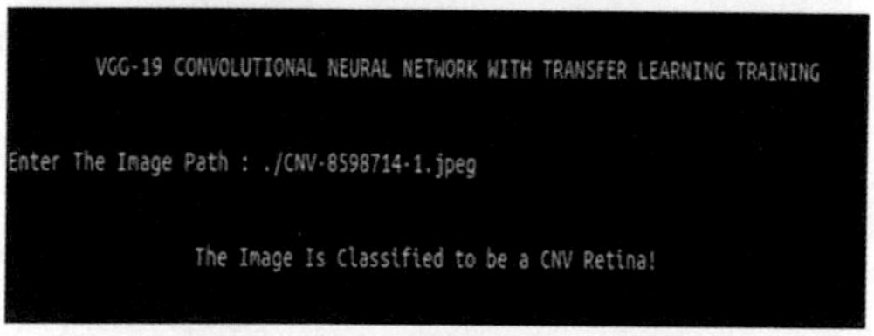

(f). VGG-19 with Transfer Training

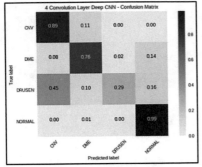

(a). 4-Convolution Layer Deep CNN

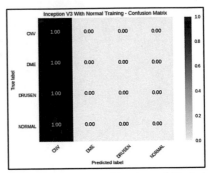

(b). Inception v3 with Normal Training

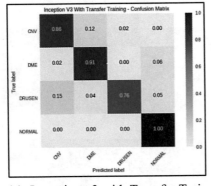

(c). Inception v3 with Transfer Training

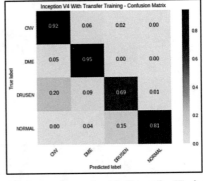

(d). Inception v4 with Transfer Training

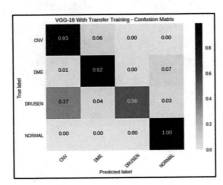

(e). VGG-16 with Transfer Training

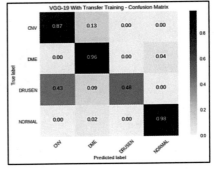

(f). VGG-19 with Transfer Training

Fig. 12 Confusion matrices for the six different architectures

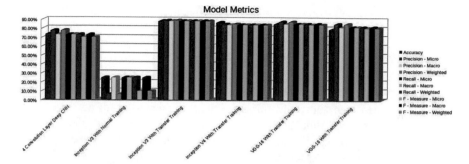

Fig. 13 Plot of model matrices for the different CNN architectures

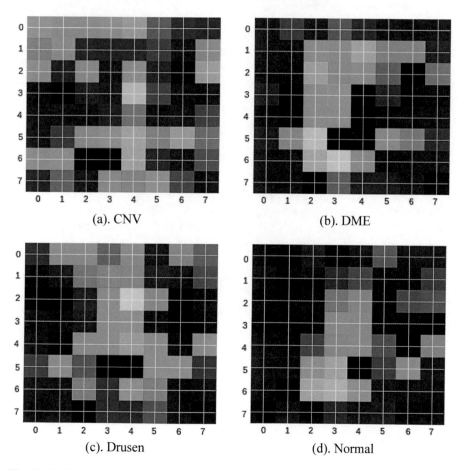

Fig. 14 Vanilla saliency maps for the retinal diseases

classified output image is shown in Fig. 11d. Thus, both Google's Inception v3 and v4 with transfer learning give better performance with superior classification abilities.

The next series of architectures with transfer training is VGG-16 and VGG-19 yielding training accuracies of 85.31% and 83.63%, respectively, with corresponding validation accuracies of 85.3% and 79.5% (shown in Figs. 9 and 10, respectively). These CNN architectures are already been pre-trained with ImageNet dataset, and thus when trained with transfer learning produces higher values of accuracy. They also require less training time due to the pre-trained ImageNet weights, and in such architectures, the learning performance is also very fast. In both cases, the amount of cross-entropy losses are less and both give perfect classified outputs when fed with OCT test images as shown in Fig. 11e, f, respectively. Thus, collectively studying CNN architectures, it is concluded that among the six architectures, Google's Inception v3 with transfer learning is recorded with the highest accuracy than Google's Inception v4 with transfer learning followed by VGG-16, VGG-19, four convolutional layer deep neural networks and Inception v3 with normal training. Moreover, architectures with transfer learning give better performance than its normal training counterpart, and hence, Google's Inception v3 and v4 architectures with transfer learning are obvious choices for the classification among the proposed ones.

4.1 Performance Parameters

To validate the performance of a system, few measures have been taken into consideration, viz. sensitivity, specificity, and accuracy. These statistical matrices are calculated in terms of true positive (TP), true negative (TN), false positive (FP), and false negative (FN) through the use of Eqs. (1)–(4).

$$Precision = TP/(TP + FP) \tag{1}$$

$$Recall = TP/(TP + FN) \tag{2}$$

$$Accuracy = (TP + TN)/(TP + TN + FP + FN) \tag{3}$$

$$FMeasure = (2 \times (Precision \times Recall))/(Precision + Recall)) \tag{4}$$

Based on true positive outcome, the system will correctly predict the presence of disease and for true negative; it correctly predicts the absence of the disease. Similarly, false positive outcome predicts the presence of the disease, whereas in reality there is no disease and false negative gives the absence of the disease in the presence of it. The way of summarization of prediction of the results classifying as TP, TN, FP, and FN is a confusion matrix. The confusion matrices depicting the performance of the different CNN architectures on a set of test data for which the true value is known are shown in Fig. 12. The matrices provide the visualization of the degree of correlation between true labels with the predicted labels for the different architectures. From the matrices, it is quite evident that for all the architectures except Google's Inception v3 with normal training, and the results obtained are satisfactory, thereby reflecting the perfect classification of the eye diseases.

The performance characteristics of the different CNN architectures are further studied in terms of precision, recall, and F-measure which are calculated from their respective test phase confusion matrices. Table 5 reports the precision achieved by different classifiers. For each classifier, the values of micro, macro, and weighted precision are found to be close to each other.

Table 6 shows the performance analysis in terms of recall. The table establishes the ingenuity of "Inception V3 with transfer training" as it achieved the highest recall. Finally, Table 7 reports a comparative analysis in terms of F-measure. In terms of F-measure, the performance of "Inception V4 (with transfer training)" and "VGG-16 (with transfer training)" is similar. However, the performance of "Inception V3 (with transfer training)" is again better than other architectures under the current study.

A plot of model matrices for the different CNN architectures is depicted in the form of a bar graph as shown in Fig. 13.

As a test to facilitate the classification process, the vanilla saliency map for the four retinal diseases under consideration is obtained as shown in Fig. 14. Thus, any slight variation in input OCT images would lead to small variation in the output and

Table 5 Performance analysis in terms of precision

Models	Precision		
	Micro (%)	Macro (%)	Weighted (%)
Four convolutional layer deep CNNs	77.74	73.10	77.74
Inception v3 with normal training	6.25	25.00	6.25
Inception v3 with transfer training	88.95	88.40	88.95
Inception v4 with transfer training	85.05	84.30	85.05
VGG-16 with transfer training	87.77	85.30	87.77
VGG-19 with transfer training	85.43	82.00	85.43

Table 6 Performance analysis in terms of recall

Models	Recall		
	Micro (%)	Macro (%)	Weighted (%)
Four convolutional layer deep CNNs	73.10	73.10	73.10
Inception v3 with normal training	25.00	25.00	25.00
Inception v3 with transfer training	88.40	88.40	88.40
Inception v4 with transfer training	84.30	84.30	84.30
VGG-16 with transfer training	85.30	85.30	85.30
VGG-19 with transfer training	82.00	82.00	82.00

Table 7 Performance analysis in terms of *F*-measure

Models	*F*-measure		
	Micro (%)	Macro (%)	Weighted (%)
Four convolutional layer deep CNNs	70.10	73.10	70.10
Inception v3 with normal training	10.00	25.00	10.00
Inception v3 with transfer training	88.25	88.40	88.25
Inception v4 with transfer training	84.14	84.30	84.14
VGG-16 with transfer training	84.60	85.30	84.60
VGG-19 with transfer training	80.94	82.00	80.94

these gradients would highlight salient image regions that most contribute toward the output.

To make the decision for the final stage classification, occlusion test is a very efficient technique. Gradient-weighted class activation mapping (Grad-CAM) technique is used in the present work to produce these occlusion maps. The occlusion maps for the retinal diseased images, shown in Fig. 15, eminently show the regions of interest of the target OCT image. These regions are the ones where the deformations have occurred for the diseased retina, and also, those in the case of a normal retina are clearly visible from the occlusion maps for the four categories of OCT images. Occlusion maps are computed over the last convolutional layer, whereas the saliency maps are computed over the dense output layers. Occlusion maps contain more details than saliency maps since they use pooling features that contain more spatial details which get lost in the dense layers.

5 Conclusion

OCT-based retinal images have been analyzed for the specific disease detection. The OCT retinal images being the input data is implemented with six different CNN models and classified results are obtained at the output. The output is correctly classified for the different four classes of the retinal images as shown in Fig. 11. The proposed model with and without pre-trained dataset has given good accuracy percentages except with the model Inception v3 without transfer learning. Hence, from the present work it can be concluded that models with transfer learning yield more accuracy than without it. At the output layers of the CNN architecture how the output is classified based on the region of interest are also shown by the saliency maps and occlusion maps for the four categories. These occlusion maps show the accurate regions of the affected retina and hence can also be shared with the clinical professionals. Thus, the method used in the present work is able to classify among the diseases as well as detect it with a lesser number of epochs and with higher accuracy. This ability of the proposed work can be used to assist the clinicians in the future where the number of patients is huge to detect the diseases. The research work

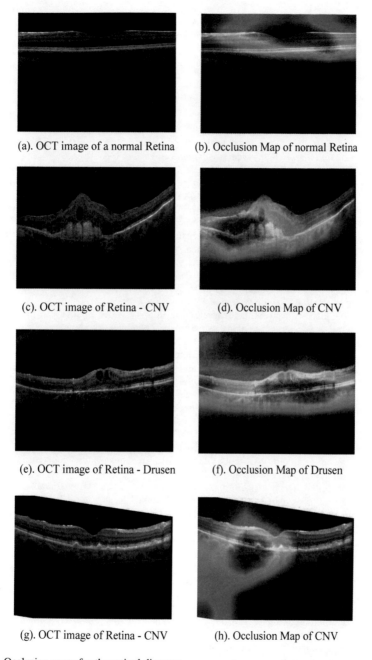

(a). OCT image of a normal Retina   (b). Occlusion Map of normal Retina

(c). OCT image of Retina - CNV   (d). Occlusion Map of CNV

(e). OCT image of Retina - Drusen   (f). Occlusion Map of Drusen

(g). OCT image of Retina - CNV   (h). Occlusion Map of CNV

Fig. 15 Occlusion maps for the retinal diseases

would definitely open up a new window of research among the research community worldwide to carry out further researches in fields of biomedical imaging.

References

1. Hassan, T., Akram, M. U., Hassan, B., Nasim, A., & Bazaz, S. A. (2015, September). Review of OCT and fundus images for detection of Macular Edema. In *2015 IEEE International Conference on Imaging Systems and Techniques (IST)* (pp. 1–4). IEEE.
2. Bagci, A. M., Ansari, R., & Shahidi, M. (2007, November). A method for detection of retinal layers by optical coherence tomography image segmentation. In *2007 IEEE/NIH Life Science Systems and Applications Workshop* (pp. 144–147). IEEE.
3. Fercher, A. F., Hitzenberger, C. K., Drexler, W., Kamp, G., & Sattmann, H. (1993). In-vivo optical coherence tomography. *American Journal of Ophthalmology, 116,* 113–115.
4. Swanson, E. A., Izatt, J. A., Hee, M. R., Huang, D., Lin, C. P., Schuman, J. S., et al. (1993). In-vivo retinal imaging by optical coherence tomography. *Optics Letters, 18,* 1864–1866.
5. Fercher, A. F. (1996). Optical coherence tomography. *Journal of Biomedical Optics, 1*(2), 157–174.
6. Regar, E., Schaar, J. A., Mont, E., Virmani, R., & Serruys, P. W. (2003). Optical coherence tomography. *Cardiovascular Radiation Medicine, 4*(4), 198–204.
7. Fujimoto, J. G., Brezinski, M. E., Tearney, G. J., Boppart, S. A., Bouma, B., Hee, M. R., et al. (1995). Optical biopsy and imaging using optical coherence tomography. *Nature Medicine, 1*(9), 970–972.
8. Bowd, C., Zangwill, L. M., Berry, C. C., Blumenthal, E. Z., Vasile, C., Sanchez-Galeana, C., et al. (2001). Detecting early glaucoma by assessment of retinal nerve fiber layer thickness and visual function. *Investigative Ophthalmology & Visual Science, 42*(9), 1993–2003.
9. Bowd, C., Zangwill, L. M., Blumenthal, E. Z., Vasile, C., Boehm, A. G., Gokhale, P. A., et al. (2002). Imaging of the optic disc and retinal nerve fiber layer: the effects of age, optic disc area, refractive error, and gender. *JOSA A, 19*(1), 197–207.
10. Otani, T., Kishi, S., & Maruyama, Y. (1999). Patterns of diabetic macular edema with optical coherence tomography. *American Journal of Ophthalmology, 127*(6), 688–693.
11. Drexler, W., & Fujimoto, J. G. (2008). State-of-the-art retinal optical coherence tomography. *Progress in Retinal and Eye Research, 27*(1), 45–88.
12. Bourne, R. R. A., Jonas, J. B., Bron, A. M., Cicinelli, M. V., Das, A., Flaxman, S. R., et al. (2018). Vision loss expert group of the global burden of disease study. Prevalence and causes of vision loss in high-income countries and in Eastern and Central Europe in 2015: magnitude, temporal trends and projections. *British Journal of Ophthalmology, 102,* 575–585.
13. Romero-Aroca, P. (2013). Current status in diabetic macular edema treatments. *World Journal of Diabetes, 4*(5), 165.
14. Rickman, C. B., Farsiu, S., Toth, C. A., & Klingeborn, M. (2013). Dry age-related macular degeneration: mechanisms, therapeutic targets, and imaging. *Investigative Ophthalmology & Visual Science, 54*(14), ORSF68–ORSF80.
15. Sengar, N., Dutta, M. K., Burget, R., & Povoda, L. (2015, July). Detection of diabetic macular edema in retinal images using a region based method. In *2015 38th International Conference on Telecommunications and Signal Processing (TSP)* (pp. 412–415). IEEE.
16. Sugmk, J., Kiattisin, S., & Leelasantitham, A. (2014, November). Automated classification between age-related macular degeneration and diabetic macular edema in OCT image using image segmentation. In *The 7th 2014 biomedical engineering international conference* (pp. 1–4). IEEE.
17. Quellec, G., Lee, K., Dolejsi, M., Garvin, M. K., Abramoff, M. D., & Sonka, M. (2010). Three-dimensional analysis of retinal layer texture: identification of fluid-filled regions in SD-OCT of the macula. *IEEE Transactions on Medical Imaging, 29*(6), 1321–1330.

18. Naz, S., Ahmed, A., Akram, M. U., & Khan, S. A. (2016, December). Automated segmentation of RPE layer for the detection of age macular degeneration using OCT images. In *2016 Sixth International Conference on Image Processing Theory, Tools and Applications (IPTA)* (pp. 1–4). IEEE.
19. Xiang, D., Tian, H., Yang, X., Shi, F., Zhu, W., Chen, H., et al. (2018). Automatic segmentation of retinal layer in OCT images with choroidal neovascularization. *IEEE Transactions on Image Processing, 27*(12), 5880–5891.
20. Parvathi, S. S., & Devi, N. (2007, December). Automatic drusen detection from colour retinal images. In *International Conference on Computational Intelligence and Multimedia Applications (ICCIMA 2007)* (Vol. 2, pp. 377–381). IEEE.
21. Zheng, Y., Wang, H., Wu, J., Gao, J., & Gee, J. C. (2011, March). Multiscale analysis revisited: Detection of drusen and vessel in digital retinal images. In *2011 IEEE International Symposium on Biomedical Imaging: From Nano to Macro* (pp. 689–692). IEEE.
22. Sharif Razavian, A., Azizpour, H., Sullivan, J., & Carlsson, S. (2014). CNN features off-the-shelf: An astounding baseline for recognition. In *Proceedings of the IEEE conference on computer vision and pattern recognition workshops* (pp. 806–813).
23. Wang, Y., Zhang, Y., Yao, Z., Zhao, R., & Zhou, F. (2016). Machine learning based detection of age-related macular degeneration (AMD) and diabetic macular edema (DME) from optical coherence tomography (OCT) images. *Biomedical Optics Express, 7*(12), 4928–4940.
24. Al-Bander, B., Al-Nuaimy, W., Al-Taee, M. A., Williams, B. M., & Zheng, Y. (2016). Diabetic macular edema grading based on deep neural networks.
25. Venhuizen, F. G., van Ginneken, B., van Asten, F., van Grinsven, M. J., Fauser, S., Hoyng, C. B., et al. (2017). Automated staging of age-related macular degeneration using optical coherence tomography. *Investigative Ophthalmology & Visual Science, 58*(4), 2318–2328.
26. Liu, L., Gao, S. S., Bailey, S. T., Huang, D., Li, D., & Jia, Y. (2015). Automated choroidal neovascularization detection algorithm for optical coherence tomography angiography. *Biomedical Optics Express, 6*(9), 3564–3576.
27. Xi, X., Meng, X., Yang, L., Nie, X., Yang, G., Chen, H., et al. (2019). Automated segmentation of choroidal neovascularization in optical coherence tomography images using multi-scale convolutional neural networks with structure prior. *Multimedia Systems, 25*(2), 95–102.
28. Khalid, S., Akram, M. U., Hassan, T., Jameel, A., & Khalil, T. (2018). Automated segmentation and quantification of drusen in fundus and optical coherence tomography images for detection of ARMD. *Journal of Digital Imaging, 31*(4), 464–476.
29. Lee, C. S., Baughman, D. M., & Lee, A. Y. (2017). Deep learning is effective for classifying normal versus age-related macular degeneration OCT images. *Ophthalmology Retina, 1*(4), 322–327.
30. Lee, C. S., Tyring, A. J., Deruyter, N. P., Wu, Y., Rokem, A., & Lee, A. Y. (2017). Deep-learning based, automated segmentation of macular edema in optical coherence tomography. *Biomedical optics express, 8*(7), 3440–3448.
31. Schlegl, T., et al. (2017). Fully automated detection and quantification of macular fluid in OCT using deep learning. *Ophthalmology, 125*(4), 549–558.
32. Kermany, D. S., et. al. (2018). Identifying medical diagnoses and treatable diseases by image-based deep learning. *Cell, 172*(5), 1122–1131.
33. Karri, S. P., Chakraborty, D., & Chatterjee, J. (2017). Transfer learning based classification of optical coherence tomography images with diabetic macular edema and dry age-related macular degeneration. *Biomedical optics express, 8*(2), 579–592.
34. Krizhevsky, A., Sutskever, I., & Hinton, G. E. (2012). Imagenet classification with deep convolutional neural networks. In *Advances in neural information processing systems* pp. 1097–1105.

Printed in the United States
By Bookmasters